Outset Geography 3

Simon Catling, Tim Firth, David Rowbotham

Series Editor: Simon Catling   Mapping Consultant: Jane Thake

Illustrated by Moira Chesmur, Jon Davis and Hamish Gordon

Oliver & Boyd

Contents

1	Patterns from Above	*page* 2
2	The Baker's Oven	8
3	Lawn Farm	14
4	A Trip to the Shops	18
5	Weather Study	24
6	Thornthwaite Forest	32
7	Never Eat Shredded Wheat	36
8	A Day on the Beach	38
9	Measuring the Map	44
10	Snowdonia National Park	50
11	A Place in the News	54
12	Pam Rekam in the Cycle Speed Trials	58
	Note on the series	62
	Teaching notes	62
	Word list	*inside back cover*

Patterns from above

A

Picture **A** is an **oblique aerial photograph** of part of a town. In the centre of this **oblique view** you can see a roundabout with trees and grass in the middle of it. A river flows across the roundabout from the lake in the right of the photograph. To help you find it, look at map **G** on page 5. If you follow the river you can see that it flows from the lake, under a building, then under a road and across the roundabout. It then flows under another road and continues on beside a park. Now look carefully at the rest of the photograph and see what other **features** you can find.

Some things to do

1a How many roads lead into the roundabout?

1b Look at the photographs in **B**. Three of them have been cut out from **A** and the other two are "tricks". Write down the numbers of the three cut-outs from **A**.

1c Choose one of the tall buildings in **A** and describe its shape and size. What do you think it is used for?

Now try these

1d Look for a small light blue shape in the top right-hand corner of **A**. What do you think it is?

1e Describe a journey by boat along the river from the lake to the park.

B

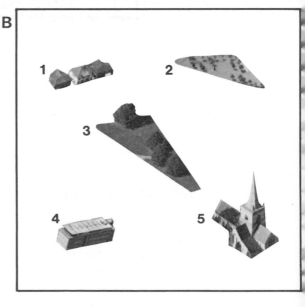

C

D

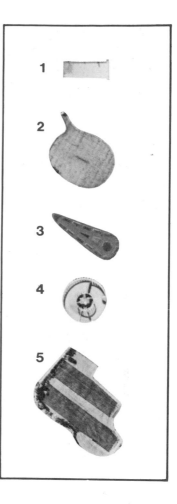

Photograph **C** is a **vertical aerial photograph** taken over the same part of town as shown in **A**. **C** shows all the features in **plan view**.

The plan view makes it easier to see the river than in **A**, but more difficult to see the buildings. Look again at the river. You will be able to see several bridges that cross it, and the shape of the lake in the right of the photograph.

Some things to do
2a How many bridges cross the river?
2b Are more or fewer than five trees growing along the banks of the river?
2c Which three cut-outs in **D** come from **C**?

Now try these
2d What was the weather like when photograph **C** was taken? What clue helped you to answer the question?
2e Photograph **C** was taken four years before **A**. Look at the big buildings near the roundabout. How can you tell that photograph **C** was taken before **A**?

E

Look at vertical aerial photograph **E**. It shows the same roundabout you saw in **A** and **C**. You can see the shape of the roundabout; it is *circular*. All the roads, buildings and spaces have their own shape. You can see the shape of these features clearly in a plan view like **E**. Some shapes are rectangles, others are squares and triangles.

Some things to do

3a Look for three different features in **E** which are circle-shaped. Write down what you think they are.

3b What shape are the gardens behind the houses?

3c Look at the houses and gardens in the bottom left-hand corner of **E**. Are most of the houses detached, semi-detached or terraced?

Now try these

3d Three of the shapes in **F** can be found in **E**. Write the numbers of these shapes and, beside each number, write the name of the feature it shows.

3e Look at the pattern formed by the roads and the river. Think of a name for this pattern, or name something it reminds you of.

F
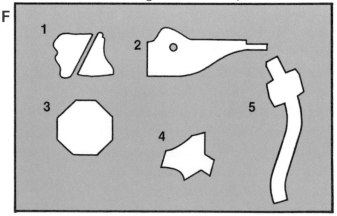

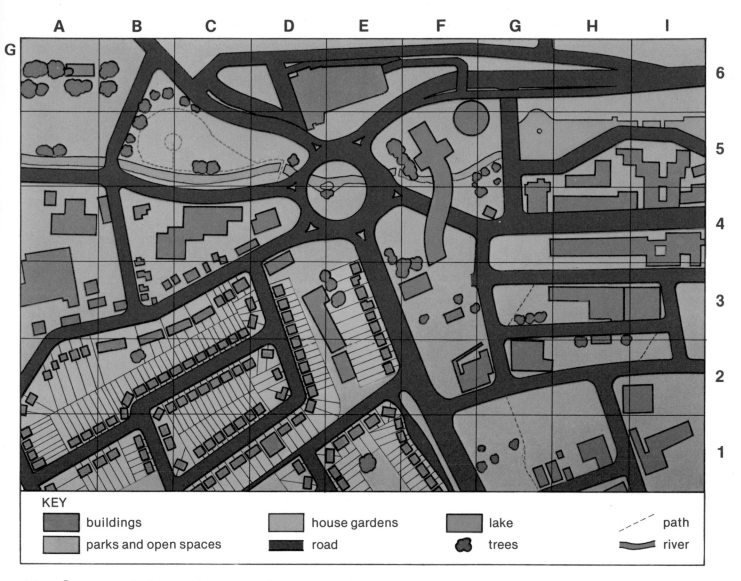

KEY

▉ buildings	▉ house gardens	▉ lake	- - - path
▉ parks and open spaces	▉ road	🌳 trees	～～ river

Map **G** was made by tracing over photograph **E**. It does not show everything that is on the ground. Some features were left out because they were too small to trace. Others were left out because there was not enough room on the map to put them in. We can say that the map shows *selected* features.

Some things to do

4a If you had to make a map of the area in photograph **E**, which features would you trace first? Why?

4b Name three features you can see in **E** which have been left off map **G**.

4c Find grid square (F, 2). Now find the same area on **E**. Does the map show everything you can see in the photograph?

Now try these

4d Eight features are shown in the key to map **G**. Write the eight names in a line across the top of a page in your workbook. Now count the number of grid squares in which you can see each of these features. Write the total number of squares under each name.

4e Look at the list you made for *4d*. Which feature did you find in most of the grid squares? Which feature was found least?

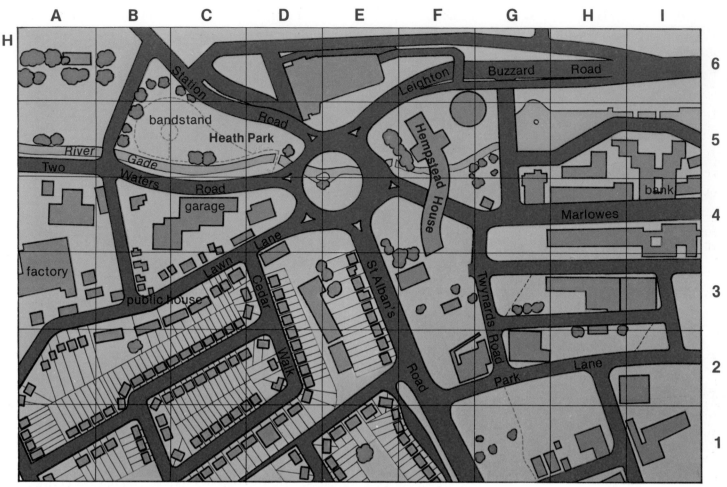

Map **H** shows the same area as map **G** but map **H** tells you the names of some of the features. It gives you more *information* about the area.

Now look again at the patterns formed by the features on the map. We can draw some of these patterns, like the ones in **I** and **J**.

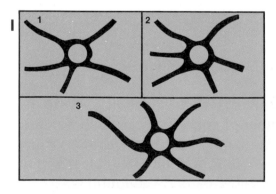

Some things to do

5a Look at **I**. Which of the three road patterns shows the roads leading into the roundabout in **H**?

5b Which pattern in **J** matches the house gardens in **H**?

Now try these

5c Trace around the edge of map **H**. Divide your tracing into four equal quarters. In each one, write the names of the features which take up most space on the ground.

5d Choose one of the four quarters and trace the pattern that the buildings make in it.

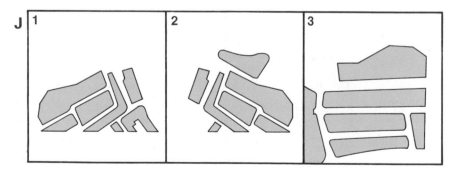

Photograph **K** shows the area at the top of photograph **C**. **L** is a map of the area in **K**. As well as showing the position of the main features in **K**, **L** also shows how the land is used. It is a **land-use map**. **L** was drawn using clues from **K** to work out what everything is. The marks on the tennis courts and on the playing fields were useful clues. So were the shapes of the buildings and the patterns they form on the ground.

Some things to do

6a Name some clues in **K** that tell us that the feature in grid square (C, 2) is a swimming pool.

6b How can you tell from **K** that part of the land in (C, 1) is used as a car park?

6c Look at the patterns of buildings in **M**. Which one fits map **L**?

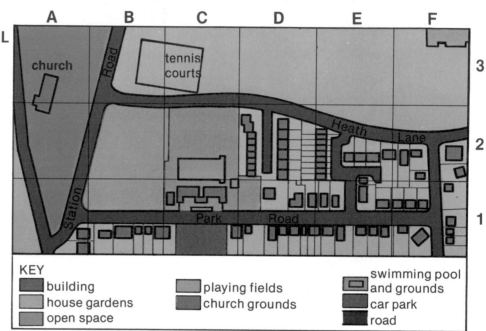

KEY

- building
- house gardens
- open space
- playing fields
- church grounds
- swimming pool and grounds
- car park
- road

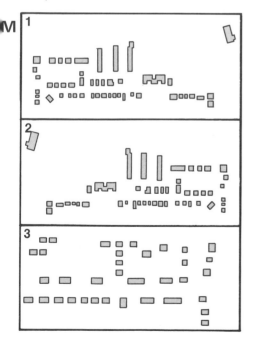

Now try these

6d Which road would you drive along to get to the swimming pool?

6e Find a clue in **K** which shows that people walk across the open space in (B, 2) on map **L**.

6f The trees you can see in **K** have been left off map **L**. Trace a copy of **L** and put the trees in the correct places.

To take you further

6g Imagine you are going to visit a town you have never been to before. To find out what the town is like, you have a choice of looking at a vertical aerial photograph or a map. Write about which one you would choose, and why.

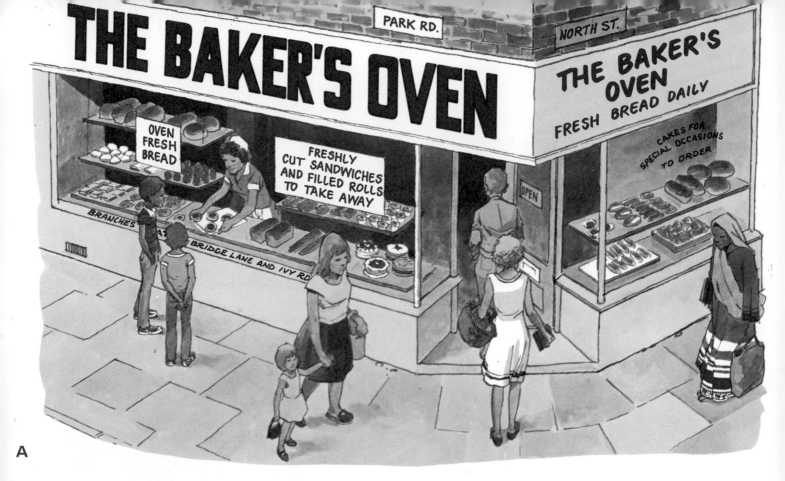

The Baker's Oven (**A**) is open every day of the week except Sunday. It sells loaves of bread, rolls and cakes. People who live and work in the neighbourhood can also buy sandwiches, hot pies and sausage rolls there at lunchtime.

The Baker's Oven is a **specialist** shop. This means that it sells *goods* of the same type, unlike a general shop which sells different types of goods. Most of the goods that are sold in the Baker's Oven are made in the bakery at the back of the shop. The bakers start work at 4 o'clock in the morning. The shop does not open until 8.00 a.m. and it closes at 5.30 p.m. The bakers finish their work at 11.00 a.m.

Some things to do
1a Name some of the goods that are sold in The Baker's Oven.
1b What type of shop is it?

1c How long have the bakers been working when the shop opens?
1d How long before the shop shuts do the bakers finish work?
1e Why do the bakers need to start work so early in the morning?

Now try these
1f The Baker's Oven is a specialist shop, so is a shoe shop. Name some others.
1g Name some examples of general shops, and write down what sort of goods they sell.
1h For how long is The Baker's Oven open each day?

Most of the goods that are sold in The Baker's Oven are made from wheat flour. Sacks of flour are delivered to the bakery three times a week. It is the end of a long journey for the flour. Some of it is made from wheat grown in Canada, and some of it comes from wheat grown in Britain.

From field to shelf

Stage 1: Growing and harvesting

The wheat which grows best in Britain is a "soft" wheat, which is good for making cakes. For breadmaking, it needs to be mixed with a "hard" wheat from Canada, which is longer lasting.

Wheat is a type of grass. It is called a **cereal** because it has a seed, or **grain**, which we can eat. Many of our breakfast cereals are made from wheat. It is the grain which is harvested and made into flour. The pictures on this page and pages 10 and 11 show what happens between harvesting the wheat and selling the bread and cakes made from wheat flour.

Some things to do

2a Name some breakfast cereals which have the word "wheat" in their names.

2b Which part of the wheat plant does the flour come from?

2c What are the main differences between the wheat fields in **B** (Canada) and those in **D** (Britain)?

Now try these

2d The ship in **C** is a grain ship. Where do you think the grain is stored on board, and how is it loaded onto the ship?

2e Think of other things you eat at breakfast. Find out what cereals they are made from.

B

C

D

Stage 2: Milling the wheat

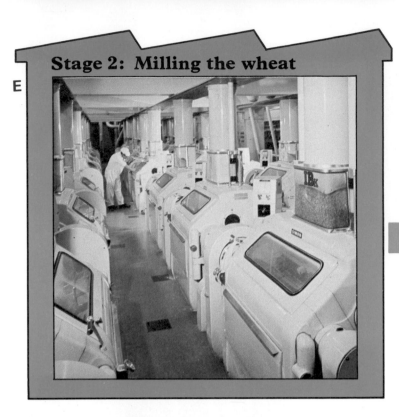

E

Flour is made by grinding, or **milling**, the wheat grain. In the mill, the grain is first cleaned and dampened to harden the outer skin, or bran. Then it is put through huge steel rollers which crush it. The crushed grain is sieved to separate the flour from the bran, and the flour is put into large sacks.

Stage 3: Baking the bread

Bread is made by mixing flour, yeast, salt and water into a dough. It is the yeast which makes the dough rise. Even in a small bakery the mixing is usually done by machine. The dough is then shaped into loaves (**F**) and put into ovens to bake (**G**).

Stage 4: Selling

H

When the loaves and rolls are taken out of the ovens (**H**), they are left to cool. Then they are put on the shelves for sale, along with the cakes, pastries and other goods (**I**).

The bread sold each day is freshly baked, and by the time the shop closes there is usually none left. Most of the bread is made with white flour, but sometimes the bran and whole wheat grains are left in to make brown "wholemeal" bread.

F

Shaping dough to make loaves and rolls, in a small bakery.

G

Loading rolls into an electric oven.

I

Some things to do

3a Name the four stages wheat goes through before we can buy a loaf of bread.

3b What happens to the grain when it is crushed?

3c How is "wholemeal" bread different from an ordinary white loaf?

Now try these

3d Describe what one of the people in the photographs is doing.

3e Not all flour is sold to bakeries. Where else might it go from the mill?

3f "Wholemeal" is one type of bread. Can you think of any others? Name them.

There are three Baker's Oven shops in town, but only the one on the corner of Park Road (**A**) has a bakery. This bakery makes the bread and cakes which are sold in the other two shops. Map **J** shows where the three **branches** of the Baker's Oven are.

The bread and cakes are delivered to the other two branches of The Baker's Oven by van. The van is loaded up at 7.30 a.m. each day and the goods are delivered in time for the shops to open at 8.30 a.m. A second delivery is made at 11.30 a.m.

Some things to do

4a Write the grid references for the three branches.

4b To make the first delivery, the van drives along Park Road and Church Road. Does it deliver to the Bridge Lane branch or the Ivy Road branch first?

Now try these

4c Describe the shortest route between the two branches.

4d Why do you think there are two deliveries a day to the two branches?

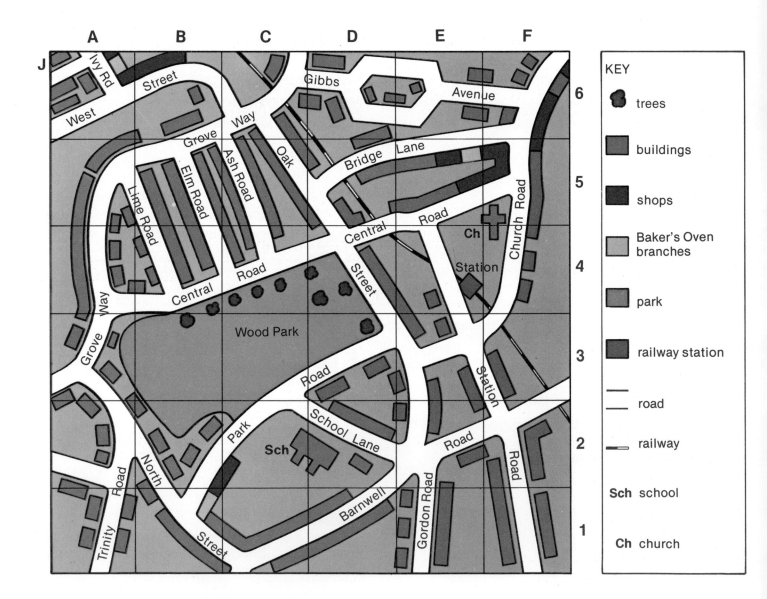

Not all the bread that is delivered to the Bridge Lane branch of The Baker's Oven is sold as loaves. Some of it is in the form of rolls, and some is made into sandwiches. Graphs **K**, **L** and **M** show some of the bread, cakes and snacks that were sold at this branch one morning. The shop was closed for a half day at 1.00 p.m. Most of the snacks were bought between midday and 1.00 p.m.

K Number of loaves and rolls sold one morning

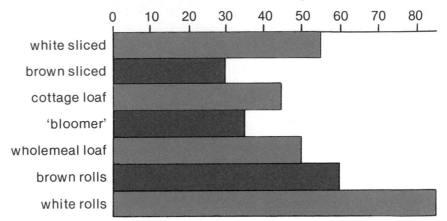

L Number of cakes sold one morning

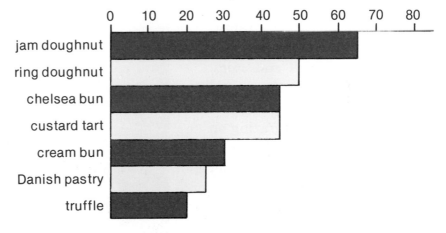

M Number of snacks sold one morning

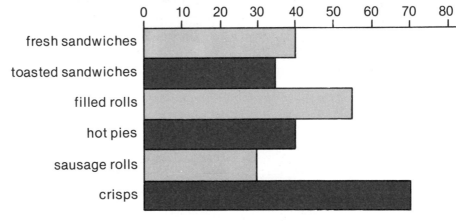

Some things to do

5a How many rolls were sold that day?

5b Were more rolls sold than loaves?

5c Which were more popular, doughnuts or buns?

5d Which kind of snack sold best?

Now try these

5e Why do you think most snacks were sold between midday and 1.00 p.m.?

5f Redraw graph **M** to show the number of snacks sold in the order of importance, starting with the highest number sold.

5g Did customers buy more cakes or more bread and rolls?

To take you further

5h Most of Canada's wheat is grown on the Canadian prairies. Find out where they are and why it is grown there.

LAWN FARM

A

Lawn Farm is in the county of Wiltshire. It is a **mixed farm**. It is called a mixed farm because it has both animals and *crops*. The farmer keeps sheep and *rears* lambs for meat. **A** shows the sheep and lambs grazing on the hillside in summer. In winter and early spring, until the lambs are born, the ewes (female sheep) are kept in large pens and barns like those in **B**. The farmer also grows grass, wheat and barley. He grows these crops to get good seed to sell to other farmers. He stores the seed in large containers called **silos** (**C**).

B

C

Some things to do
1a Why is Lawn Farm called a mixed farm?
1b Is it the same time of year in **A** and **B**?
1c What are the lambs reared for?
1d What are the silos used for?

Now try these
1e Which of the three crops grown are cereals?
1f The ewes in **B** eat hay. Where do you think it is put for them to get at easily?
1g Why do you think the ewes are kept in the pens to have their lambs?

The pictures on this page show how lambs are reared on the farm.

D

STAGE 1: THE EWES BECOME PREGNANT
RAMS FERTILISE EWES IN THE FIELDS.

STAGE 2: THE LAMBS ARE BORN AND FATTENED
THE EWES ARE BROUGHT INTO THE FARMYARD AND THE LAMBS ARE BORN UNDER COVER. THE EWES AND LAMBS GRAZE IN THE FIELDS.

STAGE 3: THE SHEEP ARE CARED FOR
SHEEP ARE SHEARED FOR THEIR WOOL.

SHEEP ARE DIPPED TO RID THEM OF BUGS.

STAGE 4: MEAT FROM THE LAMBS
THE LAMBS ARE TAKEN TO THE SLAUGHTERHOUSE.

LAMB'S MEAT IN A SUPERMARKET.

LAMB

Some things to do

2a Describe how lambs are reared.

2b What do sheep mainly feed on?

2c What do sheep provide besides meat?

Now try these

2d The sheep are dipped every year. Why is this done?

2e Why are the lambs kept on the farm for some months before they are taken away for meat?

2f Name two different sorts of shops which sell meat.

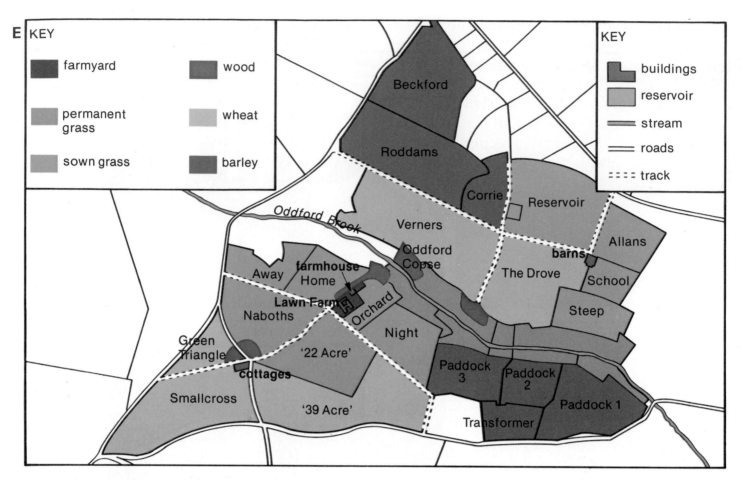

E is a **land-use map** of Lawn Farm. You can see the names of the fields, the pattern they make and what each one is used for. A stream runs through the middle of the farm. The farmer can cross this stream, but tractors have to go round by road. The fields near the stream are too wet for sheep to graze or crops to grow, so nearby farmers put their cows there to graze in summer. Every year the farmer changes the use of each field. Where wheat grows one year, sheep may graze the next.

Some things to do

3a Name the field nearest the farmhouse and the field furthest away from the barns.

3b The names of the fields are clues which tell us things about them. Choose three field names and say what you think they mean.

3c On which kind of grass do sheep graze?

3d How many fields grow wheat, how many grow barley and how many grow grass for sheep?

Now try these

3e Trace the pattern of fields on map **E**. Draw in the stream and the farmhouse.

3f On your tracing, mark the route a tractor would take from the farmhouse to Allans field.

3g Colour the fields to show how they may be used next year. Make a key to show what your colours mean.

16

The circles in **F** show all the things that are done on Lawn Farm in a year. The green circle shows how the lambs are reared. The yellow circle shows how the crops are grown. The blue circle shows how the whole farm works. The black arrows show all the things that the farm buys and the yellow arrows show all the things it sells.

Some things to do

4a What things does the farm sell:
(i) in summer, (ii) in autumn?

4b What things does the farm buy:
(i) in spring, (ii) in autumn?

4c Look at the yellow circle. Make a table with the four seasons as headings. Under each one list the jobs that are done to grow wheat, barley and grass at that time of year.

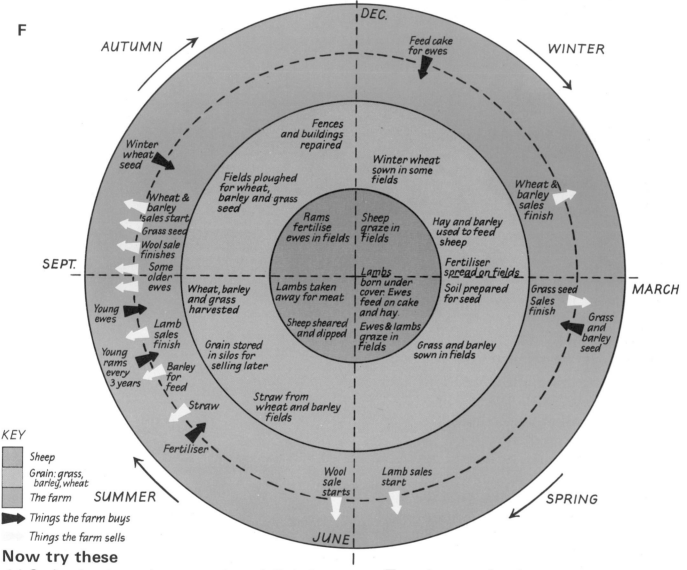

F

AUTUMN

DEC.

WINTER

Feed cake for ewes

Fences and buildings repaired

Winter wheat sown in some fields

Winter wheat seed

Wheat & barley sales finish

Fields ploughed for wheat, barley and grass seed

Wheat & barley sales start

Grass seed

Wool sale finishes

Some older ewes

Rams fertilise ewes in fields

Sheep graze in fields

Hay and barley used to feed sheep

Fertiliser spread on fields

SEPT.

Wheat, barley and grass harvested

Lambs born under cover. Ewes feed on cake and hay.

Soil prepared for seed

Grass seed Sales finish

MARCH

Young ewes

Young rams every 3 years

Lamb sales finish

Barley for feed

Grain stored in silos for selling later

Ewes & lambs graze in fields

Grass and barley sown in fields

Grass and barley seed

Straw

Straw from wheat and barley fields

Fertiliser

KEY

Sheep

Grain: grass, barley, wheat

The farm

SUMMER

Things the farm buys

Things the farm sells

Wool sale starts

Lamb sales start

SPRING

JUNE

Now try these

4d Spring is a busy time even though little is bought and sold then. Why is it so busy?

4e Straw is what is left of the barley or wheat plant after the grain has been removed. What do you think it is used for? (There is a clue in photograph **B**.)

To take you further

4f Find out all you can about either a market garden or a dairy farm. Make lists of the things that are bought and sold at different times of the year. If you can, make your lists into a diagram like **F**.

A trip to the shops

A

It is Saturday morning. Mr Weston, Sara and David have come to High Road to do the shopping and to take the clothes to the launderette. The Westons drove to the shops and parked the car behind the greengrocers. Find them in **A**.

Before they left home, the family discussed what they needed to buy and Mrs Weston wrote out a shopping list (**B**). She grouped the items together under the names of the shops where they could be bought. As well as buying all the items on the list, Mr Weston wants to collect some prints from the chemist and the children want to buy their weekly comics. When everything is done, they are going to buy fish and chips to take home for lunch.

B

Shopping List

Take washing to launderette

8 stamps	**Hardware shop**
	Screws
	Wall plugs
Greengrocer	**Supermarket**
2 mangoes	2 tins red beans
lge bunch bananas	pkt cornflakes
4 plantains	lge pkt soap powder
2 kg potatoes	3 pkts butter
1 kg carrots	pkt coffee
4 onions	pkt beef burgers
6 oranges	3 tins baked beans
	large toothpaste
Ali's	floorcloth
2 garlic	
salt fish	**Butcher**
2 kg rice	2 kg chicken
pkt chapattis	¼ kg liver
	8 sausages
Baker's	1½ kg beef joint
lge wholemeal loaf	
8 rolls	

18

When they arrived at High Road, Mr Weston checked the shopping list. He suggested that they take the laundry to the launderette first. The assistant would do the laundry for them as usual, so they could carry on with the shopping and put their bags in the car before collecting the washing. After that, they would buy the fish and chips.

When they came out of the launderette they looked at the shopping list again. David suggested they should buy the comics first, but Sara said it would be better to get them when they went to collect the laundry. She thought they should start the shopping on the other side of the road. This meant they would buy the heavy food items last, and so not have to carry heavy bags around.

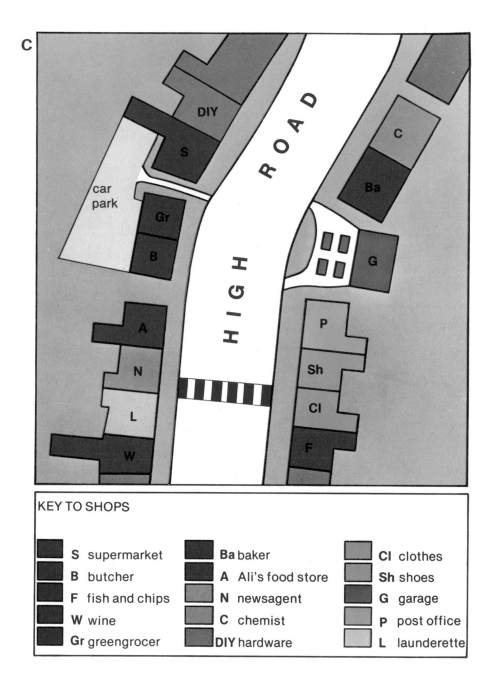

KEY TO SHOPS

S supermarket	**Ba** baker	**Cl** clothes
B butcher	**A** Ali's food store	**Sh** shoes
F fish and chips	**N** newsagent	**G** garage
W wine	**C** chemist	**P** post office
Gr greengrocer	**DIY** hardware	**L** launderette

Some things to do

1a Using **A** and **C** to help you, write down the names of the shops in front of the car park.

1b How many places will the Westons have to visit to buy the items on the shopping list?

1c Why would they buy the fish and chips after getting everything else?

1d Which three shops will the Westons visit that are not shown on the shopping list?

Now try these

1e Some shops are shown in red on map **C**. They are all the same kind of shop. What do they sell?

1f Starting with the launderette, make a list to show the order in which you think the Westons did the shopping.

1g Trace or copy map **C** and mark in the route the family took, starting at the launderette.

The shops that the Westons went to in High Road are shown in grid square (D,4) on map **D**. The Weston family live in Rose Road in the house marked in square (A,1). It is quite a long walk from Rose Road to the shops, so the Westons usually drive when they have a lot of shopping and the laundry to carry.

Most of the people living in the area shown in **D** use the High Road shops at least once a week. This is because most of the shops sell things that people need everyday, such as food and drink. A shopping centre like this one is called a **neighbourhood shopping centre**. This is because it serves the needs of the people who live in the area.

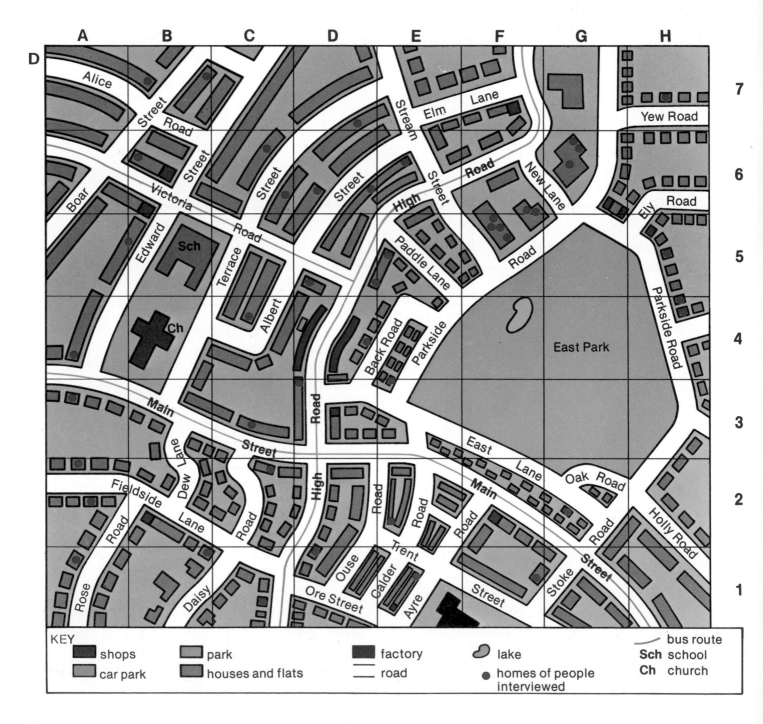

We know that the neighbourhood shopping centre in High Road serves the local area, but we do not know how big that area is. To find out, we can ask the people who use the shops where they live and how often they use the shops. We call this a **survey**. Before a survey can be done, a list of questions must be written down. This is called a **questionnaire**. E is the questionnaire used for the survey in the High Road shopping centre. Altogether, fifty people were *interviewed* for the survey. They were each asked if they would answer some questions about why they were shopping in High Road. If they said "yes", (a) to (e) were filled in on the questionnaire. Then they were asked questions 1 to 4 and the answers were written in.

Some of the information which was found out by using questionnaire **E** is shown in charts **F** and **G**. Six of the people interviewed lived in another part of the town. The homes of all the other people are marked on map **D**. The key shows you where they live.

E

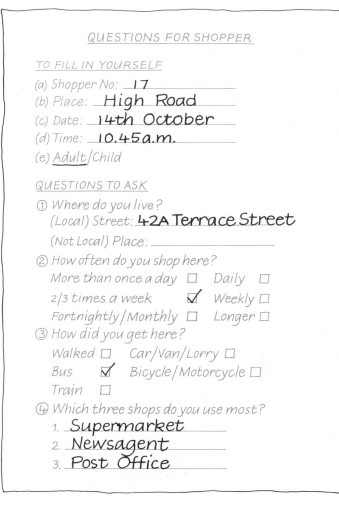

QUESTIONS FOR SHOPPER

TO FILL IN YOURSELF
(a) Shopper No: __17__
(b) Place: __High Road__
(c) Date: __14th October__
(d) Time: __10.45 a.m.__
(e) Adult/Child

QUESTIONS TO ASK
① Where do you live?
  (Local) Street: __42A Terrace Street__
  (Not Local) Place: _____
② How often do you shop here?
  More than once a day ☐   Daily ☐
  2/3 times a week ☑   Weekly ☐
  Fortnightly/Monthly ☐   Longer ☐
③ How did you get here?
  Walked ☐   Car/Van/Lorry ☐
  Bus ☑   Bicycle/Motorcycle ☐
  Train ☐
④ Which three shops do you use most?
  1. __Supermarket__
  2. __Newsagent__
  3. __Post Office__

The six most popular shops

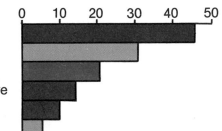

G How often shoppers came and how they came

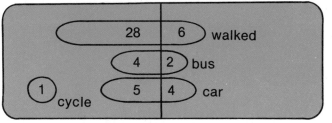

Some things to do

2a Use map **D** to work out how many of the shoppers in the survey live in the neighbourhood of the High Road shops.

2b Use **F** to find out which is the most popular shop. Why do you think this shop is used most?

2c Use **G** to find out how many people use the shops more than once a week.

Now try these

2d Why do you think three of the six most popular shops are food shops?

2e Look at map **D**. How do you think people living near the High Road shops would get to them? How might people living further away get to them?

2f Which two questions in the questionnaire do the sets in **G** use?

High Road is a very busy road. This is partly because of the shops, which attract a lot of people in cars and on foot. It is also because it is a bus route and a main road leading into the centre of town. To find out how busy it is, we can do another kind of survey. Here is what a class of children decided to do.

First, they decided to count the *traffic* that goes along High Road. They also decided to count the traffic on Victoria Road and East Lane, to see how busy they were. The children also thought they would count the number of *pedestrians* using each road.

To do their surveys, the children drew up two sheets: a **traffic count** sheet (**H**), and a **pedestrian count** sheet (**I**). They divided into three small groups and each group stood on the pavement in each of the three roads in the survey. Every time a *vehicle* or person went past them in either direction, they put a | in the correct row in their survey sheets. When they had four lines, they drew the next one across to make five, ┼┼┼.

H

PLACE: <u>High Road, outside Newsagent</u> DATE: <u>14th Oct.</u>
TIME: Start <u>10.15am.</u> Finish <u>10.30am.</u>

TRAFFIC COUNT

		Total
Cars	┼┼┼ ┼┼┼ ┼┼┼ ┼┼┼ ┼┼┼ ┼┼┼ ┼┼┼ ┼┼┼ ┼┼┼ I	
Vans and Lorries	┼┼┼ ┼┼┼ ┼┼┼ ┼┼┼ ┼┼┼ ┼┼┼ ┼┼┼ IIII	
Buses and Coaches	I	1
Bicycles and Motorcycles	┼┼┼ ┼┼┼ II	
Others	II	2

I

PLACE: <u>High Road, outside Newsagent</u> DATE: <u>14th Oct.</u>
TIME: Start <u>10.15am.</u> Finish <u>10.30a.m.</u>

PEDESTRIAN COUNT

		Total
Adults	┼┼┼ ┼┼┼ ┼┼┼ ┼┼┼ ┼┼┼ ┼┼┼ ┼┼┼ ┼┼┼ ┼┼┼ ┼┼┼ ┼┼┼ ┼┼┼ III	
Children	┼┼┼ ┼┼┼ I	11

Some things to do

3a Look at **H**. How many cars went along High Road?

3b How many adults did the children count in **I**?

3c Were the surveys in **H** and **I** done in the morning or the afternoon? How long did each one take?

Now try these

3d What was the total number of vehicles counted in High Road?

3e Were more vehicles counted than pedestrians?

3f Name one type of vehicle you would include in the row marked "others" in **H**.

When the traffic and pedestrian surveys had been completed, the children drew *block graphs* to show the information they had collected for each road (**J**). The map in **J** shows where the surveys were taken.

Some things to do

4a How many pedestrians walked along Victoria Road?

4b Did more or fewer pedestrians walk along East Lane than High Road?

4c Along which of the three roads did most cars drive?

4d Which street is the busiest?

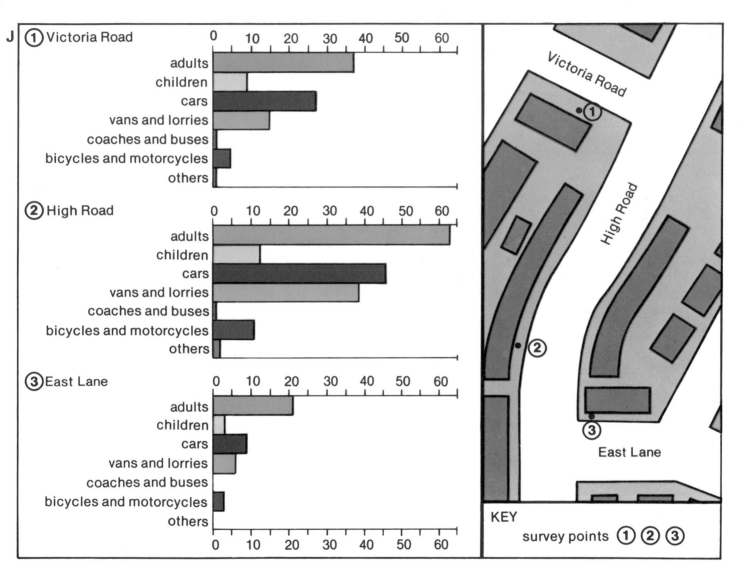

Now try these

4e Why do you think High Road is busier than the other two roads?

4f East Lane is a quieter road than the other two. There is a clue in the block graph which might tell you why this is so. What is it?

To take you further

4g Make a survey of your class to find out which local shops their families use. Make up a questionnaire to use in your survey.

Weather Study

John and Ann each made a chart to show what the weather was like on Monday. John drew chart **A** and Ann drew chart **B**. John's chart shows that it was cool on Monday. The word "cool" does not tell us very much. It can mean different things to different people. John's chart also tells us that it was showery and breezy but we do not know how much rain fell or how strong the wind was.

Ann's chart (**B**) is different from **A** because it gives us actual **measurements** for the temperature, cloud, rainfall and wind. It also uses special **symbols** which give us a more *accurate* picture of the weather than the ones that John used.

A Weather	Monday	Key
Was it hot or cold?	◑	Cool
Was it cloudy or sunny?	☁	Sun and cloud
Did it rain?	☁	Showery
Was it windy?	●→	Breezy

B Weather		Monday	Key
What was the temperature?		13 °C	13 degrees Centigrade
Was it sunny?		Some	Some sunshine
Was there any cloud?	Type	Cu	Cumulus cloud
	Cover	◕	Some cloud
Did it rain?	Type	▽	Showers
	Amount	2mm	2 millimetres
Was it windy?	Speed	⌐o	Gentle breeze
	Direction	N W	Wind from the north west

Some things to do

1a How much rain fell on Monday?

1b What does chart **B** tell you about the wind that chart **A** does not?

1c What does chart **B** tell you about the cloud that chart **A** does not?

Now try these

1d What would be the best clothes to wear if you were outside on the Monday shown in charts **A** and **B**? Write why you would choose those clothes.

1e Write down three ways in which Ann's chart is more useful than John's in showing us what the weather was like.

Temperature

Ann used special **instruments** to measure the weather for her chart. To find out what **temperature** it was, she used a thermometer like the one in **C**. The thermometer hung on a wall outside and Ann went to check it once each hour. She wrote down the temperatures in chart **D**.

C

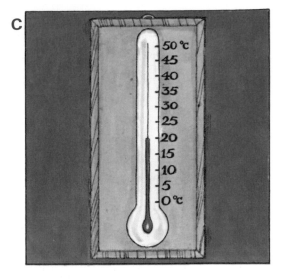

D

8 °C	10 °C	11 °C	12 °C	12 °C	13 °C	12 °C
9 a.m.	10 a.m.	11 a.m.	12 a.m.	1 p.m.	2 p.m.	3 p.m.

Ann measured the temperature every day for one school week. She then made a graph (**E**) to show the highest temperature recorded for each day. She also made a chart to show how sunny it was each day. She chose the right word from the list in **F** to put on her chart **G**.

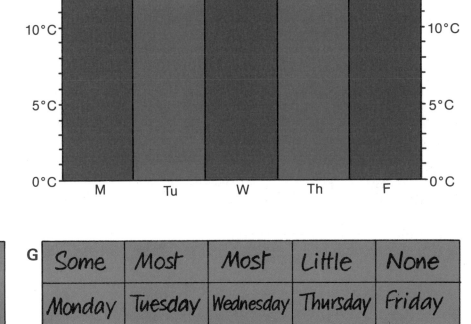

F

All Most Some Little None

G

Some	Most	Most	Little	None
Monday	Tuesday	Wednesday	Thursday	Friday

Some things to do

2a Look at chart **D**. At what time of day was the temperature highest?

2b Which days of the week had the same highest temperature?

2c What was the temperature on the day of the week when there was no sunshine?

Now try these

2d Work out the *average* temperature for the week shown in **E**. Do this by adding all the highest temperatures together, and then dividing the total by the number of days.

2e Why did Ann have to use a thermometer on an outside wall to measure the air temperature?

Cloud

Ann does not need an instrument to measure cloud, but she does need to use special symbols to show how much cloud there is in the sky. Chart **H** shows these symbols and tells you what each one means.

H

Clear sky	Little cloud	Some cloud	Mostly cloudy	Sky covered

Ann also wants to describe what type of clouds there are in the sky and to put their name and symbol on her chart. The photographs in **I** show the four main cloud types, and the names and symbols for each one.

I
1 Cirrus Ci

2 Stratus St

3 Cumulus Cu

4 Cumulonimbus Cb

Ann kept a record of the type and amount of cloud each day for a school week. **J** is her completed chart.

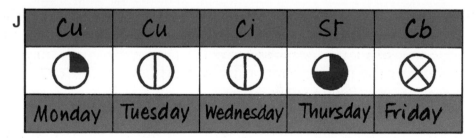

J

Cu	Cu	Ci	St	Cb
Monday	Tuesday	Wednesday	Thursday	Friday

Some things to do

3a Look at chart **J** and write out in full the names of the cloud types for each day of the week.

3b Which two days in the week shown in **J** had the same amount of cloud? Describe how much cloud there was.

3c On which day of the week was the sky completely covered with cloud?

Now try these

3d Look at the four cloud photographs in **I** and describe in your own words what each one shows.

3e Draw your own chart to show the type and amount of cloud for today.

RAIN

Ann needs two things to help her to record and measure rainfall. She needs chart **K** which gives her the symbols for three types of rain and for snow, hail and sleet. She also needs a **rain gauge** to measure how much rain falls. **L** shows what a rain gauge looks like inside. The writing at the side tells you how to make a rain gauge for yourself.

K

• Rain	✳ Snow
𝟿 Drizzle	⊛ Sleet
▽ Shower	△ Hail

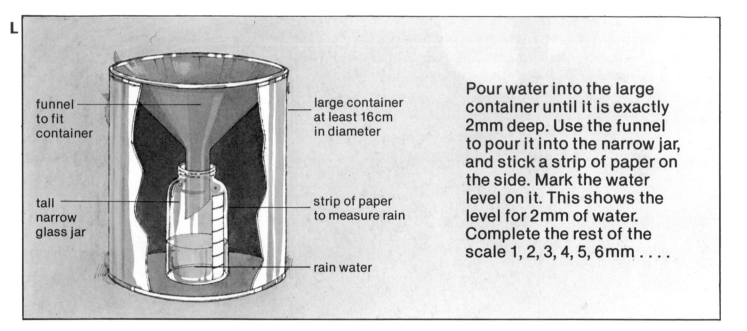

L

funnel to fit container

large container at least 16cm in diameter

tall narrow glass jar

strip of paper to measure rain

rain water

Pour water into the large container until it is exactly 2mm deep. Use the funnel to pour it into the narrow jar, and stick a strip of paper on the side. Mark the water level on it. This shows the level for 2mm of water. Complete the rest of the scale 1, 2, 3, 4, 5, 6mm . . . .

On each school day for a week, Ann checked the rain gauge to see how much water had collected. She marked the amount on a chart and added the correct symbols from **K**. **M** is her completed chart for a week.

M

▽	▽		𝟿	•
Monday	Tuesday	Wednesday	Thursday	Friday

10 mm
5 mm
0 mm

Some things to do

4a Look at the symbol for sleet in **K**. Why is this a good symbol for sleet?

4b If it snowed, how would you measure the amount that fell?

4c Which day of the week had the highest rainfall? Why was it higher than on other days?

Now try these

4d How much more rain fell on Friday than on Monday?

4e What is the best kind of place to put a rain gauge? Give reasons for your answer.

4f Why is it better to use a narrow glass jar in a rain gauge than a short wide one?

Wind

There are two things Ann wants to know about the wind. She wants to know how hard it is blowing, and where it is blowing from.

To find out how hard the wind is blowing, Ann can walk outside and guess or *estimate* the strength of the wind. She can then look at the pictures in **N** and find the symbol that best describes the wind on that day.

Ann can make a more accurate measurement of the wind strength by using an instrument called an **anemometer**. Picture **O** shows you what this looks like and how it works. Notice that the wind symbols on the anemometer are the same as those in **N**.

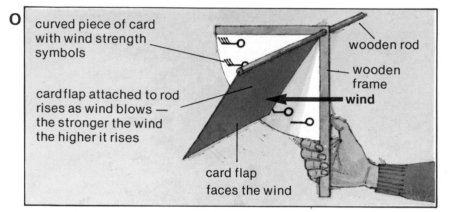

O
curved piece of card with wind strength symbols

wooden rod

wooden frame

wind

card flap attached to rod rises as wind blows — the stronger the wind the higher it rises

card flap faces the wind

P is the wind chart that Ann made for one school week.

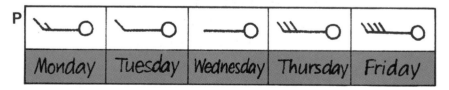

Monday	Tuesday	Wednesday	Thursday	Friday

Some things to do

5a Describe how each picture in **N** shows how hard the wind is blowing.

5b On which day of the week shown in **P** was there a gentle breeze? Which day had a strong wind?

5c If the card flap of the anemometer did not move at all, what wind symbol would you draw? Draw it and say what it means.

N calm

gentle breeze

windy

strong wind

gale

Now try these

5d Write two reasons why you think the anemometer is a better way of measuring the wind than using the pictures in **N**.

5e Draw a picture of your own to describe a strong wind.

Ann needs another instrument to find out which **direction** the wind is blowing from. It is called a **compass**. You can see one in **Q**. The needle on a compass always points towards north.

Ann stood in the playground and put the compass on the ground. She then used chalk to draw the eight points of the compass. Next, she placed a wind vane on the ground in the middle of the compass she had drawn. She made the wind vane herself. You can see it in **S** and in **R**.

Q

R

S

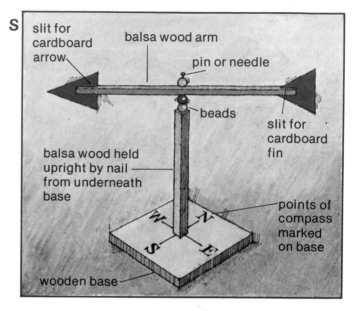

slit for cardboard arrow

balsa wood arm

pin or needle

beads

slit for cardboard fin

balsa wood held upright by nail from underneath base

points of compass marked on base

wooden base

The arrow on the wind vane points in the direction that the wind is coming from. Ann checked the wind vane each day for a week and put the directions onto her chart (**T**).

T

NW	N	–	S	SW
Monday	Tuesday	Wednesday	Thursday	Friday

Some things to do
6a Write out the eight compass directions in full. Start like this: N = north.
6b What direction is the wind blowing from in picture **R**?
6c Why is there no wind direction marked in for Wednesday in chart **T**?

Now try these
6d Why do you think that most of the wind vanes we see are on top of churches or other tall buildings?
6e If the arrow on the wind vane swings between south and east, what direction will Ann put on her chart?

On Friday afternoon, Ann collected together the weather records she had kept for the week. She put all the information on a new chart and made a key for it. Her completed record for the week is the first of the three weeks shown in **U**.

WEEK 1

U	Weather		M	Tu	W	Th	F
	What was the temperature?		13°C	12°C	14°C	13°C	13°C
	Was it sunny?		Some	most	most	little	none
	Was there any cloud?	Type	Cu	Cu	Ci	St	Cb
		Cover	◖	◑	◑	◕	⊗
	Did it rain?	Type	▽	▽	—	𝟫	●
		Amount	2mm	1mm	none	3mm	6mm
	Was it windy?	Speed	⌐○	⌐○	—○	⫣○	⫣○
		Direction	NW	N	—	S	SW

KEY

Weather		Symbol
Temperature		in °C ; degrees Centigrade
Sun		all most some little none
Cloud	Type	Cumulus Cu, Stratus St, Cirrus Ci, Cumulonimbus Cb
	Cover	clear sky ○ little cloud ◑ some cloud ◕ mostly cloudy ◕ sky covered ⊗
Rain	Type	rain ● drizzle 𝟫 shower ▽ snow ✳ sleet ✶ hail △
	Amount	in mm ; millimetres
Wind	Speed	calm ⌐○ gentle breeze ⌐○ windy ⫣○ strong wind ⫢○ gale ⫤○
	Direction	N NE E SE S SW W NW North East South West

Some things to do

7a Look at **U**. How many days had no sunshine at all?

7b Which of the three weeks had the coldest temperatures?

7c Which week had the most rain?

7d What were the temperatures on the days that had no wind?

Now try these

7e In your own words, describe what the weather was like on Friday in week 1.

7f On the days when there was no rain, what type of cloud was in the sky?

7g On the two days with the highest rainfall, which direction was the wind blowing from?

The symbols that Ann used in **U** are rather like parts of a code. We need to **interpret** the symbols before we can work out what they show. Ann was asked by her teacher to interpret the first week of her chart, and to tell the class what the weather had been like. Here is what she said.

"At the beginning of the week, there was some wind and sunshine, but it was not very warm. Some fluffy white clouds gave showers early on, but the week ended with heavy rain pouring from cloudy grey skies as a gale force wind blew. Wednesday was the warmest day and it was still, dry and sunny."

WEEK 2

M	Tu	W	Th	F
12 °C	11 °C	13 °C	15 °C	16 °C
some	none	little	most	all
St	Cb	St	Ci	—
◔	⊗	◑	◐	○
🌧	△	•	—	—
3mm	2mm	4mm	none	none
⤸o	⤸o	⤸o	⤸o	—o
SE	E	SW	SE	—

WEEK 3

M	Tu	W	Th	F
13 °C	12 °C	12 °C	10 °C	9 °C
most	some	none	none	little
Ci	Cu	St	St	Cu
◐	◔	⊗	⊗	◑
—	▽	•	•	🌧
none	1mm	7mm	5mm	2mm
⤸o	⤸o	⤸o	⤸o	⤸o
SE	S	SW	W	NW

Some things to do

8a Which day had the best weather in the second week?

8b Which day out of all three weeks had the worst weather?

8c Try to interpret the information for week 2 in chart **U**. Describe what the weather was like in your own words.

Now try these

8d Using the key to chart **U**, draw a chart which shows the symbols for a cold windy day with sunshine and showers.

8e In what ways was the weather in week 3 different from the weather in week 2?

8f **V** shows a day from chart **U**. The temperature was 13 °C. Which day was it?

To take you further

8g Imagine the weather for a week and make a chart to record it. Underneath the chart write down what the weather was like in that week.

V

A

Thornthwaite Forest

Photograph **A** shows part of Thornthwaite Forest in the Lake District of England. The forest is very large and covers many hillsides and valleys. The trees you can see in **A** are *conifers*. These trees have needle-like leaves, like the spruce tree in **B**. In other parts of the forest there are also *broadleaved* trees, such as oak which you can see in **B** too. All conifers except larch are *evergreens*. This means that they have green leaves all year round. The oak and other broadleaved trees are *deciduous*, which means they lose their leaves every autumn.

B

spruce: for hilly and wet places

oak: for lower-lying places

Some things to do
1a Look at **A**. Are most of the trees growing on the hillsides or in the valley?

1b Write down three ways in which conifers are different from broadleaves.

1c Find out the names of other conifers and broadleaves.

Now try these
1d Look at **A**. Write about the forest. Say something about the size and colour of the trees and the shape of the forest.

1e Photograph **A** was taken in August. Would it look the same in winter? Give reasons for your answer.

Map **C** shows the forest in **A**. It is divided into separate parts called **plantations**. Each one has a name. You can see one of these plantations in **D**. The plantations are separated from one another by roads, rivers and a lake. Lakes like this one give the Lake District its name. The forest is rather like a huge farm that produces wood. Plantations are planned and planted by people just like crops on a farm. The main difference between trees and farm crops is that trees take many years to grow.

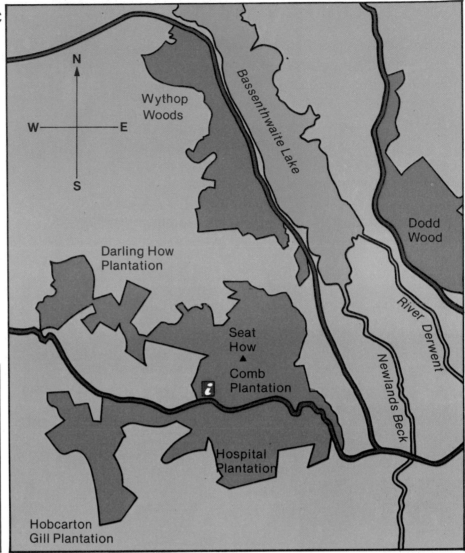

KEY

■	Thornthwaite Forest
■	lake
∿	river
▬	road
i	information centre

D

Some things to do

2a What is a plantation?

2b What is the name of the plantation in **D**?

2c In what way is a forest like a farm?

2d Why do you think the Information Centre is near the road?

2e If you were driving south past Dodd Wood, would Bassenthwaite Lake be on your left or on your right?

Now try these

2f Look at the compass directions on map **C**. Which way did the photographer face to take photograph **D**?

2g Photograph **A** was taken from the hilltop of Seat How, facing south. Which plantations does it show?

The life of a forest

Before the trees can be planted, the land has to be well *ploughed* (**E**). Young seedlings are then planted by hand about two metres apart (**F**). As the trees grow, they need more space and so some are cut down or "thinned" (**G**). This "thinning" is usually done when the trees are about 20 to 25 years old. Other trees are left to grow much bigger before they are cut down, or *felled* (**H**). The tree trunks are loaded onto a lorry (**I**) and taken to a sawmill, where they are sawn into planks (**J**).

E

F

G

H

I

J

Some things to do

3a Describe the jobs that are being done in **E** and **F**.

3b Why are forests "thinned"?

3c Make a list of things in your classroom that are made from wood.

Now try these

3d What do you think the young trees that are cut down could be used for?

3e Look at **H**. Why do you think a large piece has been cut out of the trunk before it is sawn in half?

3f Felling trees is a dangerous job. What do people wear to protect themselves?

Visiting a forest

A forest provides many jobs for people. It is also a place where people can go to enjoy themselves. In many forests there are picnic sites like the one in **K**, and paths to walk along through the trees (**L**). Information about the trees and plants is sometimes given along the route. People can also go pony trekking in forests, like the people in **M**.

Many people visit the forests, but some are not so careful as others. Fires like the one in **N** are easily started, especially in summer when everything is very dry. A fire like this can destroy many thousands of trees in a very short time.

Some things to do

4a Describe one of the activities in **K**, **L** or **M**.

4b Why do you think there are special picnic sites made in forests?

Now try these

4c How do you think the fire in **N** might have started?

4d Fires spread very quickly. Look back to **A**. You can see long breaks in the forest. What do you think these are for?

To take you further

4e Make a list of things people who visit a forest should do and not do to make sure that it is kept safe and tidy.

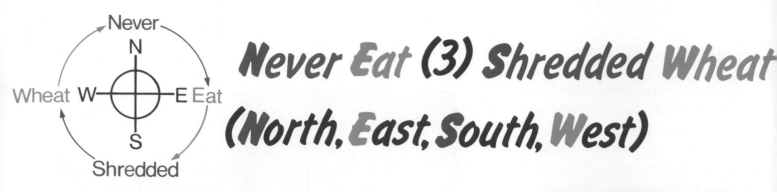

Never Eat (3) Shredded Wheat (North, East, South, West)

In picture **A**, Uzma is standing outside the school looking at the garage. We can describe where the garage is in three ways. Uzma is looking at the garage, so we can say the garage is *in front of* Uzma. Because there is a road between Uzma and the garage, we can say that the garage is *on the other side* of the road from her. The **compass directions** on the wind vane in **A** tell us that the garage is *north* of Uzma. It is therefore north of the school as well.

C is a map of the area in **A**. It does not show the wind vane but the compass directions have been written at the sides of the map to help you remember.

Some things to do

1a In **A**, is Uzma facing the school or looking away from it?

1b Look at Uzma in **B**. Is the garage in front of her or behind her?

1c Look at **B**. In which compass direction is the garage from the school?

1d On map **C**, are the bushes east or west of the school?

Now try these

1e In **A**, are the flower beds to the east or west of Uzma?

1f In **B**, Uzma has turned round. Are the flower beds still in the same compass direction from Uzma as they were in **A**?

1g In **B**, if Uzma turned towards the flower beds, which compass direction would be behind her?

A

B

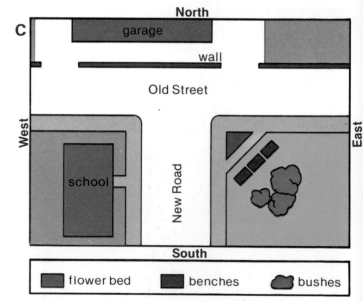

C

North

garage

wall

Old Street

West

East

New Road

school

South

flower bed · benches · bushes

The wind vane that you saw in **A** and **B** is fixed in position. This is because compass directions are fixed. They never change. When Uzma turned round, the garage was still north of the school but Uzma was facing south.

Now look at map **D**. It is the same as map **H** on page 6, but it has been made smaller and has been turned round. On map **D**, the lake is in the top left-hand corner, but in map **H** on page 6, the lake was in the top right-hand corner. There were no compass directions marked on map **H** on page 6 so we did not know if the lake was north, east, south or west of the roundabout. By using the compass directions on map **D**, we can work out that the lake is north of the roundabout.

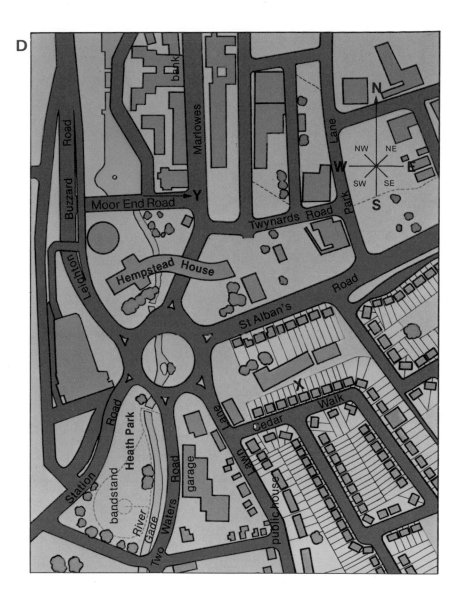

Some things to do

2a Is the River Gade on the east side or the west side of map **D**?

2b Name two roads that lead into the roundabout from the south.

2c Is the park north or south of the roundabout?

2d To drive east from the roundabout, which road would you go along?

Now try these

2e Are most of the small buildings in the south-west or the south-east of map **D**?

2f If you walked from the lake towards point Y on map **D**, which direction would you come from?

2g Look back to map **H** on page 6. Now fill in the blanks in this sentence with the names of the correct compass directions. "The right-hand side of the map is the ... side and the bottom of the map is the ... side. The roundabout is ... of the lake."

To take you further

2h Imagine you live in the house marked X on map **D**. Name the streets you would walk along to get to the lake, and give the compass direction for each turn you make.

A DAY ON THE BEACH

A

The weather had been hot and sunny during the week. Jane and Peter Watts decided that if it stayed fine on Saturday, they would take their three children to the seaside for the day. The weather was lovely so the family drove to the coast on the Saturday morning. They set off early so that they would have a long day on the beach and so that they could get there before it became too crowded.

B

Because the weather was so good, hundreds of other people decided to go to the seaside for the day. People who do this are called **day trippers**. On the Saturday morning the roads to the coast were full of cars (**A**) and the trains were packed with passengers (**B**). It is on days like this that you may hear news reports like the one in **C**. You usually hear them during the months of June, July and August.

C There were traffic jams again today as day trippers headed for the beaches while the hot weather lasts. By 9 o'clock this morning, traffic was very heavy on all main roads, and journeys back tonight will be slow. Trains have also been packed with passengers going to the coast.

Some things to do
1a Name two ways in which people travel to the seaside.

1b Draw two sets like the ones below and write three words from this page that go in each set.

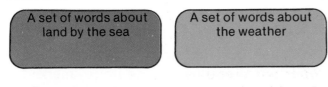
A set of words about land by the sea   A set of words about the weather

1c Describe what you can see in either **A** or **B**.

1d What is a day tripper?

Now try these
1e Why do day trippers leave early in the morning to reach the coast?

1f Read the report in **C**. Why will journeys be slow in the evening?

1g Why are June, July and August the most popular months for people to go to the seaside?

Most of the day trippers and **holiday makers** who go to the coast go to a **seaside resort**. This is a special type of town where there are plenty of **facilities** for visitors to enjoy. These facilities include shops (**E**) which sell things that people like to buy when they are on holiday, such as postcards, beach balls, buckets and spades and swimming costumes. As well as the shops, there are usually lots of cafés and restaurants, and parks where games like tennis and clock golf (**F**) can be played. Some resorts also have a pier like the one in **G** and **H**. Piers often have amusement arcades or theatres at the end and seats where people can sit and enjoy the view.

D

E

F

G

H

Some things to do

2a Look at **D**. Name three things that the people are doing on the beach.

2b Name four facilities you might find at a seaside resort.

2c Name six things that you might want to buy in a seaside resort.

Now try these

2d What is a seaside resort?

2e Name some things you could do if you spent a day at a seaside resort.

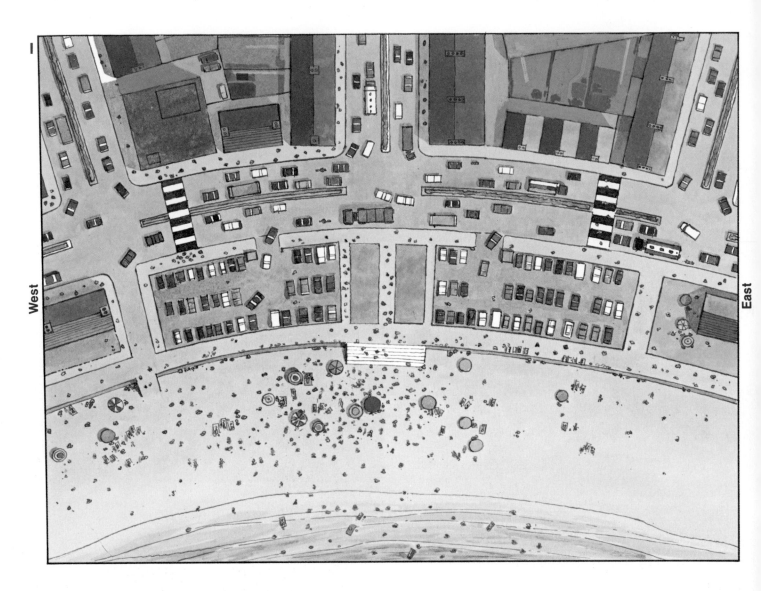

West

East

I is a plan view of the beach and part of the resort where the Watts family went. They arrived at 9.30 a.m. and parked the car in the eastern car park. By 11 o'clock both car parks were full and the beach was crowded. I shows what the beach looked like at 11 a.m. You can find the Watts family in I because they sat under a red sunshade.

Some things to do
3a Is the middle, the east end or the west end of the beach most crowded?

3b Trace or draw a map of I. Remember not to put in people or cars, and give the map a key and a north sign.

3c An ice cream van arrives at the beach and wants to park nearest to where most people are on the beach. On your map, draw a X where you think the van should park.

Now try these
3d Explain why you put your X for the ice cream van where you did.

3e Why do you think that the beach is more crowded in the middle and the west than the east side? There are clues on I to help you.

3f What do you think the buildings nearest to the beach might be? Give a reason for your answer.

Day trippers are not the only people who visit seaside resorts. There are also holiday makers who stay in hotels (**J**) or guest houses (**K**) for several days or even weeks. Other visitors may rent holiday flats or caravans.

When the weather is good, most holiday makers spend their time on the beach, but if the weather is poor, they look for other things to do indoors. As well as shopping in the bigger shops near the centre of the resort (**L**), they can go to amusement centres, play bingo (**M**), go to cinemas, theatres or to restaurants. Most of these places are open in the evenings.

Some things to do

4a Name four sorts of place where holiday makers can stay in seaside resorts.

4b What is the difference between a day tripper and a holiday maker?

4c Make a chart like the one below. In each column write down the names of places that visitors can go to. Use all the photographs to help you. You can put the names of some places in more than one column.

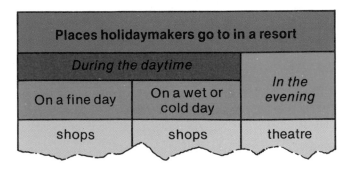

Places holidaymakers go to in a resort		
During the daytime		*In the evening*
On a fine day	On a wet or cold day	
shops	shops	theatre

Now try these

4d Look at the chart you made for question 4c. Choose three of the places you listed and explain why you put them where you did in your chart.

4e Of all the places that you listed in your chart, which ones, if any, would you not find in an ordinary town?

41

As you have seen, there are a lot of places for holiday makers to go to in seaside resorts. All these places need people to run them or to work in them so there are plenty of jobs for people to do. **N** shows some of these jobs. You will find others in the photographs you have looked at.

Because most visitors to resorts go there in the summer, this is the busiest time of the year. There are very few visitors at other times of the year. This means that in the summer, more people are needed to work in the shops, hotels, theatres and so on. Many of them live in the resort, but others come in from outside in the summer.

N

Some things to do

5a Make a copy of the chart below. In the correct column, list the jobs shown in **N**. Now add the names of other jobs shown in the photographs on pages 39–41.

Jobs done in the holiday season	Jobs done all year round
deckchair attendant	bus driver

5b How many jobs on your chart are done only in the summer holiday season?

5c Choose one holiday season job and explain why you put it in that list.

Now try these

5d Many of the people who come to work in resorts in the summer are students. Why do you think this is?

5e How are workers who stay at resorts to work during the holiday season like the holiday makers and day trippers?

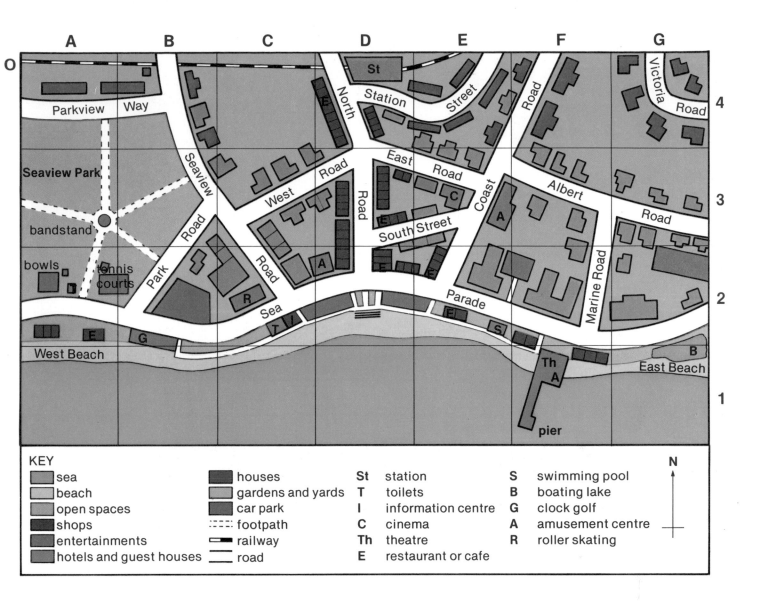

At a seaside resort, you usually find most of the big hotels and many of the shops and entertainment facilities near to the sea. **O** is a land-use map of a resort. It shows how the land and buildings are used in the part of the resort nearest to the beach.

Some things to do

6a Name six entertainment facilities you can see in **O**. Use the key to help you.

6b Look carefully at rows 2, 3 and 4 of the grid in map **O**. In which row are there most entertainment facilities?

6c In which row are there most hotels and guest houses?

6d Which road is the main shopping street?

Now try these

6e Why do you think many of the large hotels were built facing the sea?

6f Why are there so many shops on the road between the station and the beach?

To take you further

6g Imagine you went to spend a week's holiday in the resort shown in **O**. Write about how you would get there and how you would spend your time.

Measuring the Map

Look at **A**. It is a vertical photograph of the corner of a table. The whole table is one metre long and half a metre wide. Because the table is so much bigger than this book, it is impossible to show all of it at its actual size.

Now look at **B**. This shows the whole table, but the table looks much smaller than it actually is. The photograph shows the table at a *reduced size*. We can **reduce** the photograph to even smaller sizes, as shown in **C**.

A

Some things to do

1a Look at **B**. Name the three objects on the table.

1b Is the book in **B** shown at its actual size?

1c When something is reduced in size, is it made smaller or larger?

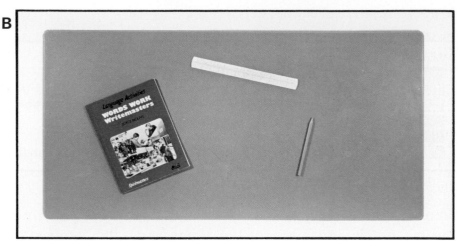

B

Now try these

1d If photographs were taken of the objects on the table in **B**, which ones would have to be reduced to fit them onto a page in this book?

1e Trace the outline of the table in **C2**. Now trace the outline of **C3** so that it fits into the bottom left-hand corner of your tracing of **C2**. How much bigger is **C2** than **C3**?

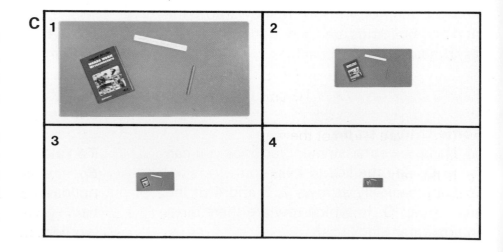

C 1   2   3   4

D is a plan of the table shown in **A**, **B** and **C**. Like photograph **B**, it shows the table at a smaller size than it really is. **D** has been drawn accurately, so that the table in the plan is exactly ten times smaller than the real table. This means that a distance of 1 centimetre (cm) on the plan is equal to 10 centimetres on the real table.

E shows part of a ruler with centimetres marked on it. Look at the ruler next to plan **D**; you can see that the table in **D** is exactly 10 cm long. We know that each centimetre on the plan equals 10 cm on the real table, so the length of the real table is: 10×10 cm $= 100$ cm.

When a plan is drawn accurately like **D**, we say it has been **drawn to scale**. The scale is:

1 cm on the plan $=$ 10 cm real size.

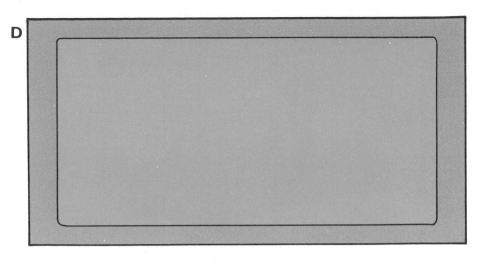

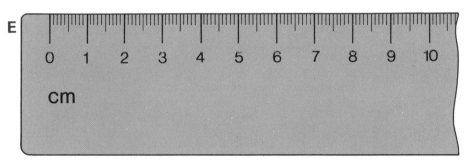

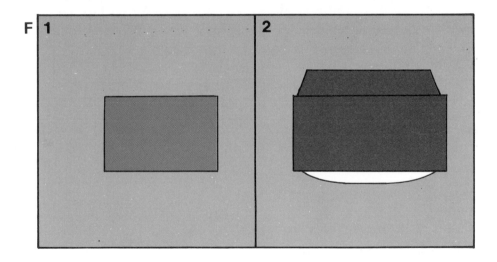

Some things to do
2a You know that the length of the table on plan **D** is 10 cm. Now measure the width.

2b What is the width of the real table?

2c Is the table shown in **B** the same size as the table shown in plan **D**? Use your ruler to measure it.

Now try these
In **F**, 1 cm on the plans = 10 cm on the real objects.

2d Use the scale to work out the width of the real atlas shown in plan **F1**.

2e How long is the real television shown in plan **F2**?

2f Measure the table or desk you are sitting at. Using a scale of 1 cm on the plan = 10 cm real size, draw a plan to scale to show your table.

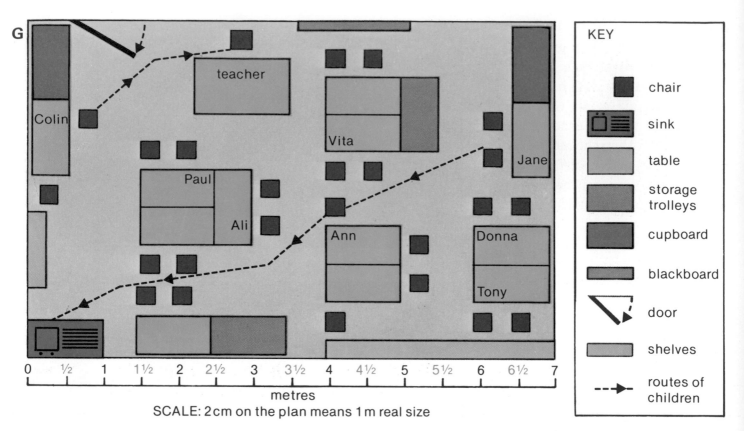

G

KEY

◼ chair

▦ sink

▭ table

▭ storage trolleys

◼ cupboard

▭ blackboard

◢ door

▭ shelves

--▶ routes of children

```
0   ½   1   1½   2   2½   3   3½   4   4½   5   5½   6   6½   7
```
metres

SCALE: 2cm on the plan means 1m real size

G is a plan of a classroom that has been drawn to scale. You can work out the size of the actual classroom by using the **scale bar** under plan **G**. The scale bar is like a ruler. You can use it to measure distances on the plan. On this scale bar 2 cm are equal to 1 metre (m) on the ground:

2 cm on the plan = 1 m real size.

Pictures **H** and **I** show you how to use the scale bar. First, take a piece of paper and lay the edge of it along the line on the plan that you want to measure. Now mark the paper at each end of the line you are measuring. Put your paper against the scale bar so that the first mark is under the 0 on the scale. Now look at the scale to see where the second mark on your paper touches it. In **I** you can see that the distance from the front of Ali's desk to the wall is 3 metres.

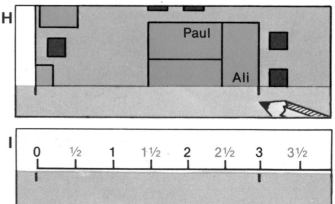

H

Paul

Ali

I

```
0   ½   1   1½   2   2½   3   3½
```

Some things to do

3a Use a piece of paper and the scale bar in **G** to work out the length of the classroom in metres, from left to right.

3b What is the width of the classroom (**G**) in metres?

3c How far is Jane's chair from Ann's chair?

Now try these

3d How long is Colin's route to the teacher?

3e How long is Ann's route to the sink?

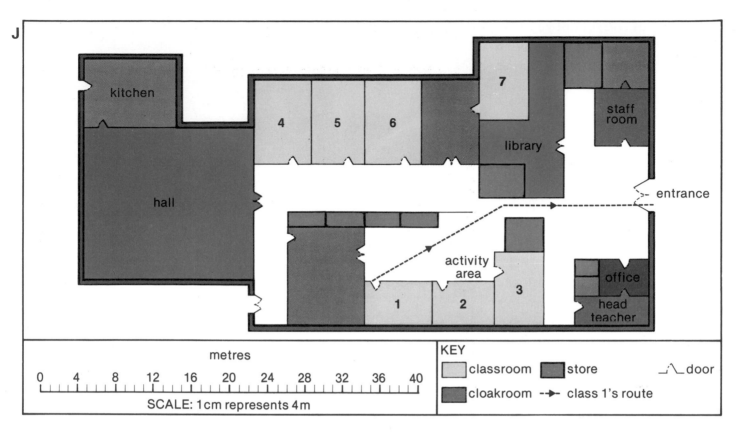

J

metres										
0	4	8	12	16	20	24	28	32	36	40

SCALE: 1cm represents 4m

KEY
☐ classroom ▮ store ＿⌐＿ door
▮ cloakroom -▸- class 1's route

J is a plan of a school. It has been drawn at a smaller scale than the classroom in **G**. The scale bar under **J** shows that 1 centimetre on the plan stands for, or *represents*, 4 m on the ground.

You can see that the school building is longer than the scale bar. To measure distances which are longer than the scale bar, this is what you should do. Lay a piece of paper along the line on the plan that you want to measure, then mark the two ends of the line. Put your first mark against the 0 on the scale bar and make another mark on your paper under the 40-metre point on the scale bar. Picture **K1** shows you how to do it. Then move your paper along so that the 40-metre mark is under the 0 on the scale. Now measure the distance between your 40-metre mark and the last mark on your paper (**K2**). If you add this distance to 40 m, you will have the full distance of the line you wanted to measure.

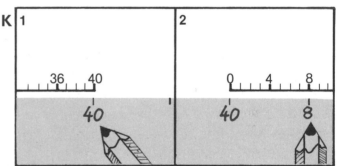

K

Some things to do

4a What is the scale of plan **J**?

4b How wide is plan **J** from left to right?

4c Measure class 1's route from their classroom to the school entrance.

Now try these

4d Measure the distance of the widest part of the school building.

4e How far is it from the door of class 6 to the hall doors?

4f Is the entrance door more or less than 50 m from the kitchen door?

L is a plan of the school you saw in J, but it is drawn to a smaller scale. You can see the same school in plan M, which is drawn at an even smaller scale. The more of the ground you want to show, the smaller the scale needs to be.

In L you can see the school playground and flower beds but the plan tells you nothing about the area round the school. In M, the scale is too small to show the flower beds, but you can see much more of the area round the school. The smaller the scale of the plan or map, the less detail you can put on it.

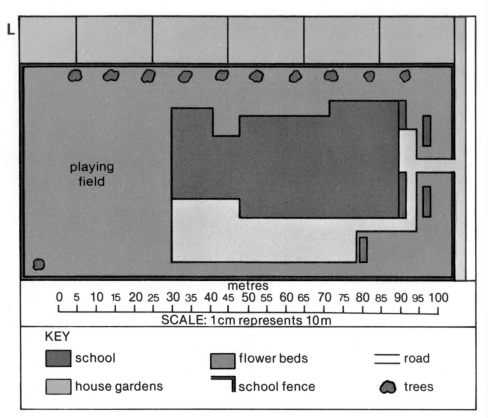

Some things to do

5a What is the scale of L and the scale of M?

5b How long is the school building in L?

5c Are the school grounds more or less than 100 m long?

5d On map M, is Broom Avenue 50 m, 100 m or 150 m long?

5e What is the width of map M?

Now try these

5f What is the distance between the trees along the top edge of L?

5g Why do you think the flower beds in L are not shown on map M?

5h Is the plan of the school in L two, four or five times bigger than it is in map M?

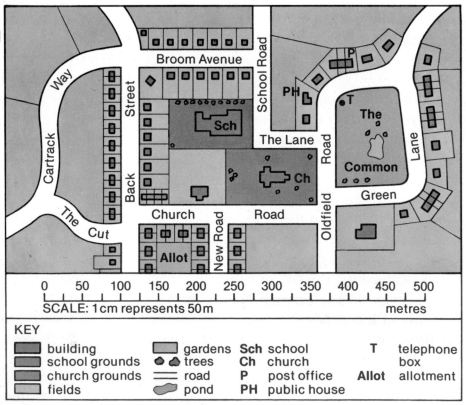

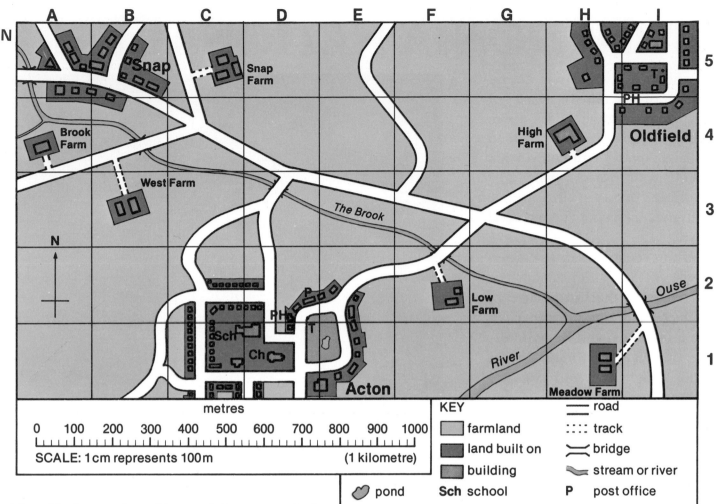

Map **N** shows the village of Acton, and the area to the north of it. This map shows a larger area at a smaller scale than map **M**. Many of the roads in **N** are curved. To measure these you can use a piece of cotton, thin string or a pipe cleaner. Bend the cotton along the length of road you want to measure, then measure the cotton against the scale bar as you did before with a piece of paper.

Some things to do

6a In which village on map **N** is the school you saw in **L** and **M**?

6b What is the straight-line distance in metres between Low Farm and West Farm?

6c How much of the River Ouse can you see on **N**? Is it 700, 800 or 900 m?

Now try these

6d Look at Acton in **M** and **N**. Which features have been left off **N** that are on **M**?

6e Measure the straight-line distance from the telephone box in (I, 5) to the telephone box in (D, 2). Now measure the distance by road. Which is the longest?

6f How far would you have to walk using tracks and roads to get from Snap Farm to Low Farm?

To take you further

6g Measure the length and width of your classroom in metres. Choose a scale then draw a plan of the classroom to scale. Put the scale underneath your plan.

SNOWDONIA NATIONAL PARK

A shows an **oblique aerial view** of a small part of Snowdonia National Park. Even this part is much larger than an ordinary park. You could walk round your nearest park in a morning, but to walk round a National Park would take several days.

You can see mountains, *valleys*, rivers and lakes. These are the **features** of the **landscape** in **A**. Snowdon and Glyder Fach are mountains, and Snowdon is the highest in England and Wales. The rivers and lakes are in the valleys.

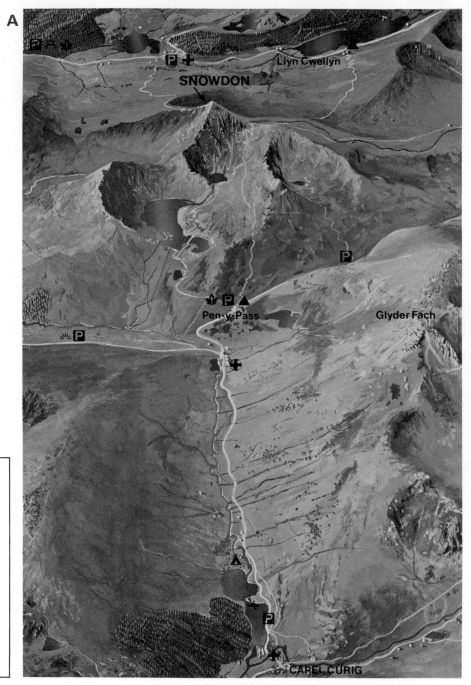

A

SNOWDON

Llyn Cwellyn

Pen-y-Pass

Glyder Fach

CAPEL CURIG

KEY TO SYMBOLS

- *i* information centre
- ✚ mountain rescue post
- ···· national park boundary
- **P** car park
- ⚑ picnic site
- ▲ youth hostel
- ✿ nature trail
- 🚐 caravan site
- ▲ camp site
- ⚲ canoeing
- ∪ pony trekking
- 🏰 castle
- 🚂 railway

Some things to do

1a What features can you see in **A**?

1b Think of three words to describe the mountains in **A**.

1c What features can you see in the bottoms of the valleys?

Now try these

1d Draw a picture of the peak of Snowdon.

1e Describe the shape which you have drawn.

1f Describe the shape of the valley which goes from Pen-y-Pass to Capel Curig.

The landscape of Snowdonia is very beautiful. People want to keep it like this, and so in 1951 it was made a National Park. There are nine other National Parks in England and Wales; Snowdonia is one of the largest. As with a local park, people come to spend their *leisure* time here. They can come for the day or for a holiday. Photographs **B** to **F** show some of the things people do when they come.

D

B

E

C

F

Some things to do

2a Describe the activities people are doing in any two of photographs **B** to **E**.

2b How long has Snowdonia been a National Park?

2c Why do we have National Parks?

2d Look at **A**. Are there more youth hostels or car parks?

Now try these

2e Look at the symbols on **A**, and the key. What do they tell you about the kinds of holiday that people have in Snowdonia?

2f What do you think has happened in **F**?

2g Describe a day's walk from the youth hostel at Pen-y-Pass, up Snowdon to the youth hostel at Llyn Cwellyn.

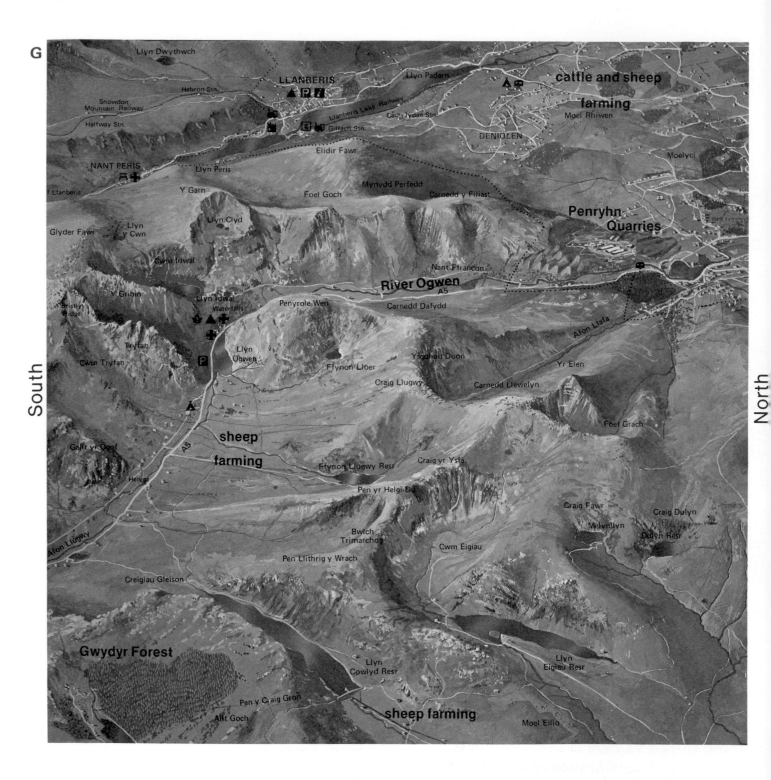

G shows the landscape to the north-east of Mount Snowdon. Part of it can also be seen in **A**. As well as showing landscape features, such as rivers and mountains, **G** also contains clues to the jobs done by people who live in the area.

One way some people in this part of Snowdonia earn their living is by farming. They farm sheep (**H**) and cattle (**I**). Other people work in the forests (**J**), and others work to keep the landscape beautiful for people to enjoy.

H

I

J

K

Another very important job in Snowdonia is slate quarrying (**K**). The slate is cut from the face of the rock. It is used mainly for roofs, but also for things such as flooring, work surfaces and billiard tables. **K** shows part of Penrhyn Quarries. Find them on **G**. Penrhyn Quarries are the largest working slate quarry in the world.

Some things to do

3a Find and name the largest forest in **G**.

3b Describe one of the jobs being done in **H**, **I**, **J** or **K**.

3c Which job involves removing part of the land?

3d What ways of earning a living are not shown here? The photographs on page 51 will give you a clue.

Now try these

3e Holiday makers and farmers both want to use the land. Describe how **E** on page 51 shows this.

3f Imagine you are standing on top of Carnedd Dafydd and facing south. Name three places you could see.

To take you further

3g Design a poster to attract tourists to Snowdonia.

3h Design and make up a pamphlet to go with your poster. Use maps and pictures in the pamphlet.

A PLACE IN THE NEWS

The pictures in **A** have three things in common. First, they are all news reports. Television, radio, newspapers and magazines are different *media* for reporting news, or things which may interest people.

A

Each report in **A** names a *place*. It may be a town or city, such as Moscow, or a country, such as Iceland. It is almost impossible to read or listen to the news without seeing or hearing the name of a place.

All the places named in **A** are in the **continent** of **Europe**. **B** is a map of Europe. It shows the names of all the countries. It also shows where the places named in **A** are.

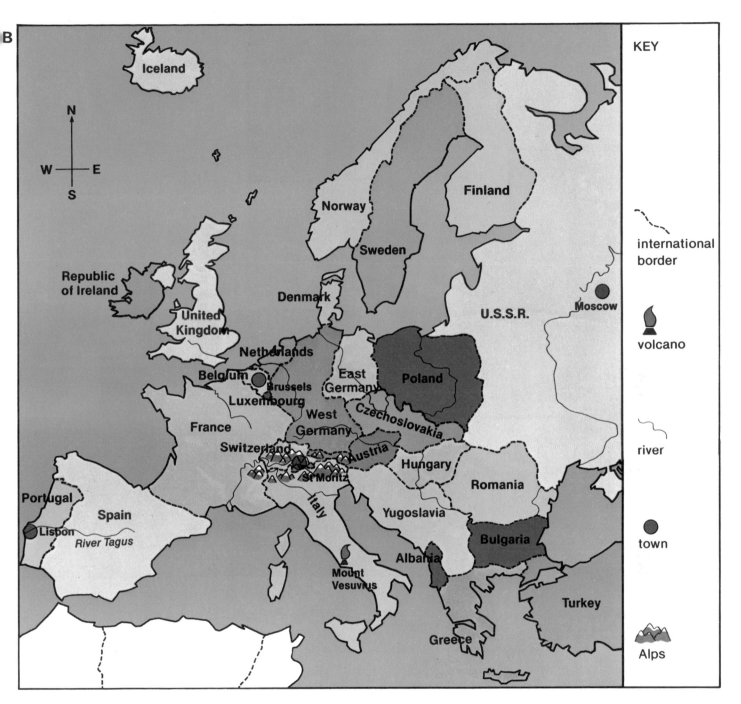

B

KEY

- - - international border

🌋 volcano

～ river

● town

⛰️ Alps

Some things to do

1a Which place named in **A** is an island?

1b Which of the places named in **A** is furthest to the east?

1c In which country is St Moritz?

1d Which two countries does the River Tagus flow through?

1e Is Mount Vesuvius in the north or south of Europe?

Now try these

1f Draw or trace map **B** of Europe.

1g The following places are in the news: Paris, the North Sea, Poland, Mont Blanc, Dublin, Sicily, Gibraltar, Helsinki. Use an *atlas* to find out where they are and mark them on your map.

Tony Collins is a newspaper reporter. He is working in Copenhagen when he gets a telephone call from his office in London. Read on, and follow his journey on map **E**

Some things to do

2a Why does Tony go to Sicily?

2b How does he travel from Copenhagen to Palermo?

2c How does he report back to his paper?

2d Imagine you are Tony. Describe your journey from Copenhagen to Mount Etna.

Now try these

2e Which type of media does Tony work for?

2f Why do you think Tony flew to Sicily, rather than going by train?

2g Why do you think the newspaper included a map with Tony's report?

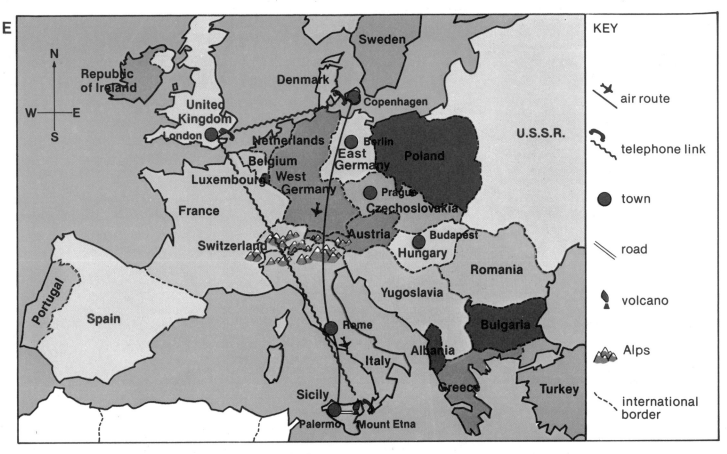

E

KEY
- ✈ air route
- ∿ telephone link
- ● town
- ⫽ road
- 🌋 volcano
- ⛰ Alps
- ‐‐‐ international border

Tony Collins reports on events in Europe for his newspaper. The people in his office need to be able to contact him quickly if they hear of a story. They do this by telephoning him. The telephone is a *link* between Tony and his office. He also needs to be able to move about Europe quickly. The aeroplane is a fast *link* between Copenhagen and Palermo. These **links** are important to Tony. They are also important to other people, for his newspaper report is the link between them and what is happening in Europe. Map **E** shows the links that made it possible for Tony to report on the eruption of Etna only a day after he left Copenhagen.

Some things to do

3a What was the link that Tony used between Copenhagen and London?

3b What was the link that Tony used between Palermo and Mount Etna?

Now try these

3c Imagine you are writing to a penfriend. What is the link between you?

3d From Sicily, Tony has to go back to Rome and then to Budapest, Prague and Berlin before returning to London. Look at map **E**. What countries are these cities in?

3e On your tracing of map **B**, mark the cities Tony goes to between Palermo and London. Draw his route and make a key to show how he travels.

3f Why are fast links important for reporters like Tony?

To take you further

3g Collect copies of some daily newspapers. Check them to see how many places in Europe are named. List the places, then make a map to show where they are.

PAM REKAM in *THE CYCLE SPEED TRIALS*

Pam is competing in the famous Forest Track Cycle Speed Trials. Competitors are timed over a course which goes through the forest and along country lanes and farm tracks. You can see the area in maps **A** and **B**. The competitors have to find out the route themselves, using clues, a compass and a map. Each cyclist starts in turn, and does not know where the finish is until the last clue is found! Follow Pam's route on **A** and **B** and watch out for her mistakes!

Pam is ready to go.

THE FOREST TRACK CYCLE SPEED TRIALS

...and Pam's off NOW!

The way to the first forest track is marked with an arrow.

At the gate, Pam gets her first clue to tell her where to go.

Here's your first clue.

She looks at her map.

It says 'East to fourth turn, then south, west, south again, but you're not out of the woods yet!' So first I go east.

Pam stops at the T-junction in (E,5).

East is to my left, so I go right here.

Clue 1. East to fourth... south... but you... the wood...

Pam arrives at the second checkpoint and collects clue 2.

Thanks. Now let's see — 'West past the pond, then south for 700 metres to a building with wood in it's name.' That must be Wood House in (D,2).

Pam reaches Wood House in (D,2).

No, not here, miss.

Pam looks at her map again.

It must be Wood Farm. I forgot to use the scale bar to check the distance.

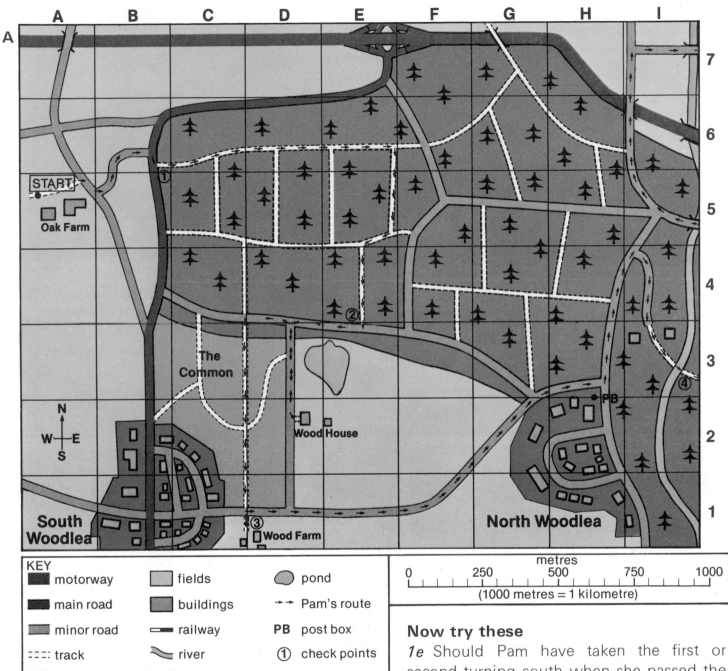

KEY

▬	motorway		fields	⬭	pond
▬	main road		buildings	•—►	Pam's route
▬	minor road	▬	railway	**PB**	post box
┈┈	track	≈	river	①	check points
▓	forest	⌣	bridge		

metres

0   250   500   750   1000

(1000 metres = 1 kilometre)

Now try these

1e Should Pam have taken the first or second turning south when she passed the pond?

1f Describe what Pam would have seen on her journey from the first to the second checkpoint.

1g Make a copy of the roads and tracks in map **A**. On your map, draw the quickest route from Oak Farm to Wood Farm.

1h Using a piece of string, measure the distance Pam cycled between checkpoints 1 and 2.

Some things to do

1a In which direction did Pam cycle first?

1b Give the grid reference for the second checkpoint.

1c Read clue 2 again. Name the building Pam should have gone to from the pond.

1d Why did Pam go to the wrong building?

Eventually, Pam reaches Wood Farm where she collects the next clue.

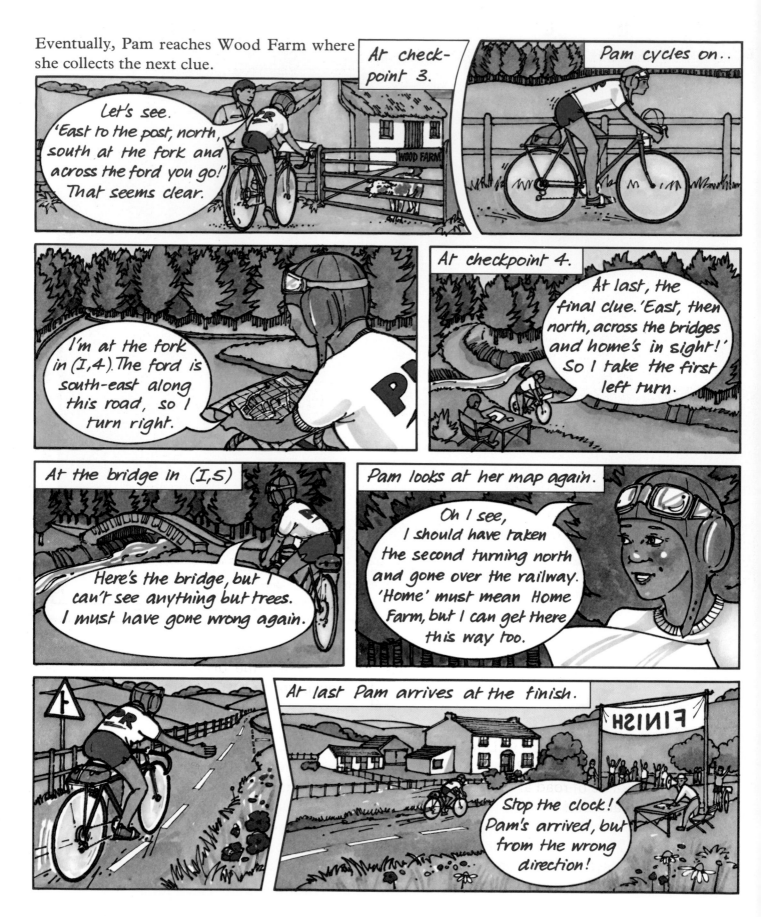

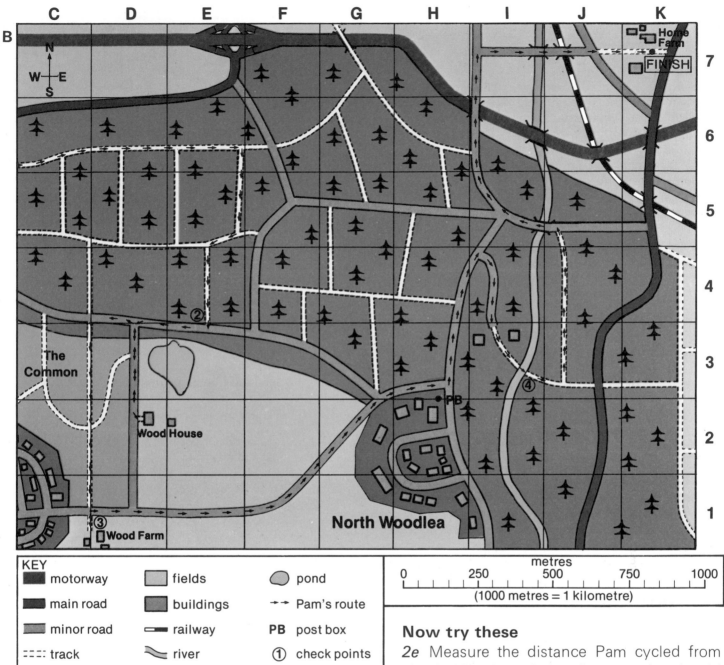

KEY

▬ motorway	▭ fields	⬭ pond
▬ main road	▬ buildings	•→ Pam's route
▬ minor road	▭ railway	**PB** post box
┅ track	〰 river	① check points
▨ forest	〓 bridge	

metres				
0	250	500	750	1000

(1000 metres = 1 kilometre)

Some things to do

2a Find the motorway on map **B**. Does it run north to south or east to west?

2b What other kinds of road are there? Draw the symbol for each and name it.

2c From which direction did Pam arrive at Home Farm?

2d How many more bridges did she cross than she should have?

Now try these

2e Measure the distance Pam cycled from checkpoint 4, and the distance she should have cycled. How much further did she go than was necessary?

2f Look at both maps. How many times did Pam turn left on her journey from the start to the finish?

To take you further

2g Write out directions for the quickest route between checkpoint 1 and Home Farm. Give the compass direction for each turn.

The Outset Geography series

In recent years there has been growing concern about the fragmented nature and limited extent of geographical and environmental work in primary schools. The H.M.I. survey, *Primary Education in England*, underlined the lack of organisation in this area of the curriculum. In response to this situation, this series has been developed to provide a framework for geographical and environmental studies for 7–12 year olds. Both the teaching methods used in primary schools and the changing nature of geography in recent years have been taken into consideration, as has the need to provide a sound foundation for developing geographical skills and understanding in secondary education.

Aims:

1 To develop oracy, literacy, graphicacy, numeracy and motor skills.
2 To introduce and develop the skills, concepts and ideas specifically related to geographical and environmental study.
3 To develop the child's understanding of places and people, and their interdependence.

Method of approach:

Outset Geography uses and builds upon the child's own experience. Geographical skills and concepts are initially introduced in contexts that are familiar to most children – the home, school and neighbourhood – in both urban and rural settings. These same themes and concepts are re-examined in more remote contexts in later books, building on what has already been learnt to give a progressive development of key ideas. Even the more remote studies – remote in terms of both distance and knowledge – are linked back into the child's own experience.

Structure and use:

Each book in the series consists of 12 units, each consisting of 1–4 double-page spreads. The progressive development enables teachers to use the series as a structured course. For those who prefer to adopt a topic or project based approach, each unit may be studied independently, and used as core material.

Special features:

Exercises: Throughout each unit there are sets of graded exercises which encourage the extraction and explanation of information from both text and illustrations. **Some things to do** require straightforward skills and should be within the reach of all children. **Now try these** are more demanding exercises, designed to stretch the quicker and more able child. Additional work which encourages further research is given at the end of each unit in **To take you further**.

Readability and vocabulary: Special attention has been given to the readability of the text and to the introduction of specialised vocabulary appropriate for the age group. Specialised words and phrases are printed in **bold**, and are listed, together with others requiring special attention, at the end of each book.

Illustrations: The 'readability' of maps and diagrams has been given special priority. Specialist advice was sought on children's perceptions of shape, colour, symbolisation, etc., and used in the preparation of the illustrations. As with the text, the maps and diagrams show a graded development throughout the series.

Book 3: Teaching notes

Book 3 develops the children's geographical understanding built up in Books 1 and 2, and extends their appreciation, both of their own experiences and of the local neighbourhood. It also fosters their awareness of the wider world, not through studies of physically distant places, but through an introduction to the wide range of homeland environments. The specific resources, ideas and skills of geographical and environmental study are extended to a level appropriate to 9–10 year old children (see the matrix on the inside front cover). As in Books 1 and 2, many of the units in this book can be developed through the personal experience of the child, either by undertaking studies in the locality (e.g. *Weather Study* and *A Trip to the Shops*), or by using the child's knowledge and experience (e.g. *A Place in the News* and *A Day on the Beach*). Several may involve both (e.g. *The Baker's Oven* and *Snowdonia*). Also included are studies which extend the child's knowledge well beyond direct experience (e.g. *Lawn Farm* and *Thornthwaite Forest*) and others which introduce more advanced map and photo interpretation skills (e.g. *Patterns from Above*, *Never Eat Shredded Wheat* and *Measuring the Map*). The following guidelines highlight the key points in each unit, and offer suggestions for additional resources (both in and out of school) and for project work.

Teaching notes

KEY POINTS	RESOURCES	Ideas for Extension Work
Unit 1 Patterns from Above 1 Oblique and vertical viewpoints and their relation to maps. 2 Identification of features from photographs. 3 Shape and size of features. 4 Patterns of features. 5 Land-use map. 6 Comparison of map and photograph.	1 Oblique and vertical aerial photographs of own area. 2 1:2500 and 1:10 000 scale maps of own area. 3 Patterns of lines and shapes linked to given maps etc. 4 Maps of different types, e.g. street, road, OS maps of large scale.	1 Look for 'local patterns' on maps and photographs of school neighbourhood. 2 Project on 'Patterns from Above' looking for patterns in urban and rural landscapes on maps and photographs and linking these to work in art and maths. 3 Make a land-use map of the local area.
Unit 2 The Baker's Oven 1 Specialist shops. 2 From wheat to bread: the stages in production. 3 Reinforcement of map grid and flow diagram. 4 The bakery as a central production point in a chain of retail branches. 5 Patterns of buying shown by graphs.	1 Information on local specialist shops, re distribution and sales. 2 Local map to show distribution of specialist shops. 3 Background information on wheat/flour production. 4 Examples of flour in different stages from seed to fine flour and bread.	1 Project on flour-based food production or a local baker or local shop which has several branches. 2 Study the pattern of sales in a local bread shop. 3 A survey of the sources of bread and cakes in the local area. 4 Simple bread or cake cookery in class.
Unit 3 Lawn Farm 1 Idea of a mixed farm. 2 Cycle of meat production from birth of lambs to sale of meat. 3 Farm land-use and field rotation. 4 The farm as a seasonal system: cycle of production, inputs and outputs, inter-dependence of activities and products.	1 Visit to an accessible farm, and/or cattle market. 2 Information on sheep farming in other parts of Britain and the world, e.g. New Zealand and Australia. 3 Information on the uses of sheep and lambs for wool. Examples of wool clothing.	1 Project on sheep farming or farms of varied types. 2 A study of what happens to the meat or wool from sheep. 3 Make a model of a farm, and discuss use of land, perhaps over several years. 4 Discuss the annual cycle, seasonally influenced, on farms.
Unit 4 A Trip to the Shops 1 Planning a shopping trip: choosing best order and route in relation to needs. 2 How to organise and conduct local surveys. 3 Idea of neighbourhood shopping centre and of catchment area of shops. 4 Study of local shoppers' habits and of traffic and pedestrian flow.	1 Base map of neighbourhood shopping centre and of potential catchment area. 2 Prepared shopping lists to use with the map. 3 Photographs of local shops. 4 Information on variety of products in local shops.	1 Comparative study of what can be bought in the local corner shop and in the neighbourhood shopping centre. 2 Make own shoppers and/or traffic surveys. 3 Class study of shopping patterns of children's families.
Unit 5 Weather Study 1 Comparison of accurate weather measurements with simple observations. 2 Use of instruments to measure the elements of weather. 3 Recording the weather accurately. 4 Compass directions. 5 Interpretation of weather records. 6 Patterns of weather over several weeks.	1 Actual weather instruments (including compass) and/or material for making own instruments. 2 Photographs of different weather conditions. 3 Visit to a local weather recording station, e.g. local secondary school, park, lighthouse. 4 Information from other sources on weather recording, e.g. link to science work.	1 Make own weather recording instruments and use these to record elements of weather over a fortnight or longer. 2 Develop language work in weather and look at reports and stories of dramatic weather conditions and effects. 3 Compare weather records kept at different times of year, and in different years.
Unit 6 Thornthwaite Forest 1 Types of trees, and the forest landscape. 2 Wood as a 'farmed' product. 3 Cycle of production: long term. 4 Forests as managed areas. 5 Recreational use of forests: some dangers.	1 Information on tree types. 2 Photographs of forest and woodland areas. 3 Information on forestry development, e.g. from Foresty Commission. 4 Visit to an accessible woodland area that is managed.	1 Project on the 'farming' and uses of wood. 2 Consider how forests affect the landscape. 3 Study the trees and their life-cycle in local streets and parks etc. 4 Plant a tree.

Unit 7   Never Eat Shredded Wheat

1 Idea of relative position developed from personal to abstract reference system.
2 Compass points fixed.
3 Compass directions in relation to maps.

1 A compass.
2 Local and other maps showing compass directions for use in orientation exercises.

1 Make up mnemonics for NESW.
2 Look for clues locally and on maps which mention compass directions, e.g. 'West Gate' and 'Southfields' etc.
3 Devise activities for children to follow compass directions.
4 Link to work on angles and magnetism.

Unit 8   A Day on the Beach

1 Special features of a seaside resort.
2 Facilities in a resort, and their location.
3 Influence of beach site.
4 Seasonal variation of jobs in a resort.
5 Influence of weather on leisure.

1 Postcards/photographs of resorts and seaside facilities.
2 Brochures of seaside resorts.
3 Maps of resorts.
4 Model of a resort.
5 Field trip to an accessible seaside resort.

1 Study resorts the children have been to.
2 Find out about specific famous seaside resorts, including their early history.
3 Design adverts for a resort.
4 Using a U.K. map, locate the main seaside resorts and nearby cities.

Unit 9   Measuring the Map

1 Plans as reduced drawings of real objects.
2 Maps and plans as scaled representations of area.
3 Using a scale bar.
4 Map content related to scale.
5 Scale in relation to area shown.

1 Scale–shape puzzles: choosing shape of appropriate scale (for map) to represent same shape at a larger scale.
2 Photos of same object at different scales.
3 Plans and maps at different scales showing e.g. desks, classroom, school, etc.
4 Scale-drawn base maps of classroom/ school on which pupils can add features drawn to scale.
5 Plans of houses from e.g. estate agents.

1 Link to measurement work in maths.
2 Draw objects and areas to scale, and at various scales.
3 Compare the content of maps of different scales.
4 Make scaled models based on maps.
5 Compare different distances by walking them. Estimate real distances.

Unit 10   Snowdonia National Park

1 Comparison between local and National parks.
2 Landscape features of National Park.
3 Interpretation of oblique view.
4 Jobs and leisure activities in a National Park.
5 Park as a conserved and managed area.

1 Visit to nearest National Park, National Trust area or area of outstanding beauty.
2 Information on National Parks, e.g. from Countryside Commission, Department of Environment and National Park Centres.
3 Photos of such areas from magazines.

1 Project on National Parks, or on idea of conservation of areas of natural beauty.
2 Study recreational activities in these areas.
3 Discuss the theme of conservation versus exploitation of land.
4 Study the location of Britain's National Parks and their remoteness or nearness to population centres.

Unit 11   A Place in the News

1 News: how it informs us about places.
2 The role of the reporter, and the transmission of news.
3 Travel and communication: choosing the most appropriate means for the circumstances.
4 The map of Europe: shape and countries.
5 Links between places.

1 Collection of newspaper and magazine news articles referring to different places.
2 Postcards, holiday brochures etc. for use in location exercises.
3 Information on travel links in Europe: air lines, ferries, motorways etc.
4 Up-to-date wall maps, route maps, atlas maps of Europe.
5 Labels of products from European locations.

1 Plot the distribution of places linked to known events in, or products from, Europe.
2 Study of links in the transport chain for goods and information moving around Europe.
3 Introduction to economic units in Europe, e.g. European Community and Warsaw Pact nations.
4 Exercises on the best types of transport/ communication to use for a variety of circumstances.

Unit 12   Pam Rekam in the Cycle Speed Trials

1 Interpretation of written clues to find routes.
2 The use of map (scale, compass directions, symbols and grid references) to find routes on the ground.
3 Interpretation of medium-scale maps.

1 Teacher-drawn copies of Pam Rekam's map for children to plot routes on.
2 Materials for children to produce own maps for orienteering activities.
3 Mazes for children to find routes through using compass directions and measured distances.
4 Local and other area street and O.S. (1:10 000) maps.

1 Plan and map a trail round the local area, and link to clues and compass directions.
2 Devise a game on 'finding the hidden casket on Treasure Island' using a base map and written directions.
3 Orienteering activity around school site, using clues.